Holger Wolfschmidt

Substrate Effects in Electrocatalysis

Holger Wolfschmidt

Substrate Effects in Electrocatalysis

Electrocatalysis from large nanostructured surfaces down to single particles - experimental and theoretical insights.

Südwestdeutscher Verlag für Hochschulschriften

Impressum/Imprint (nur für Deutschland/only for Germany)
Bibliografische Information der Deutschen Nationalbibliothek: Die Deutsche Nationalbibliothek verzeichnet diese Publikation in der Deutschen Nationalbibliografie; detaillierte bibliografische Daten sind im Internet über http://dnb.d-nb.de abrufbar.
Alle in diesem Buch genannten Marken und Produktnamen unterliegen warenzeichen-, marken- oder patentrechtlichem Schutz bzw. sind Warenzeichen oder eingetragene Warenzeichen der jeweiligen Inhaber. Die Wiedergabe von Marken, Produktnamen, Gebrauchsnamen, Handelsnamen, Warenbezeichnungen u.s.w. in diesem Werk berechtigt auch ohne besondere Kennzeichnung nicht zu der Annahme, dass solche Namen im Sinne der Warenzeichen- und Markenschutzgesetzgebung als frei zu betrachten wären und daher von jedermann benutzt werden dürften.

Coverbild: www.ingimage.com

Verlag: Südwestdeutscher Verlag für Hochschulschriften GmbH & Co. KG
Heinrich-Böcking-Str. 6-8, 66121 Saarbrücken, Deutschland
Telefon +49 681 37 20 271-1, Telefax +49 681 37 20 271-0
Email: info@svh-verlag.de

Approved by: München, TU, Diss., 2011

Herstellung in Deutschland:
Schaltungsdienst Lange o.H.G., Berlin
Books on Demand GmbH, Norderstedt
Reha GmbH, Saarbrücken
Amazon Distribution GmbH, Leipzig
ISBN: 978-3-8381-3197-9

Imprint (only for USA, GB)
Bibliographic information published by the Deutsche Nationalbibliothek: The Deutsche Nationalbibliothek lists this publication in the Deutsche Nationalbibliografie; detailed bibliographic data are available in the Internet at http://dnb.d-nb.de.
Any brand names and product names mentioned in this book are subject to trademark, brand or patent protection and are trademarks or registered trademarks of their respective holders. The use of brand names, product names, common names, trade names, product descriptions etc. even without a particular marking in this works is in no way to be construed to mean that such names may be regarded as unrestricted in respect of trademark and brand protection legislation and could thus be used by anyone.

Cover image: www.ingimage.com

Publisher: Südwestdeutscher Verlag für Hochschulschriften GmbH & Co. KG
Heinrich-Böcking-Str. 6-8, 66121 Saarbrücken, Germany
Phone +49 681 37 20 271-1, Fax +49 681 37 20 271-0
Email: info@svh-verlag.de

Printed in the U.S.A.
Printed in the U.K. by (see last page)
ISBN: 978-3-8381-3197-9

Copyright © 2012 by the author and Südwestdeutscher Verlag für Hochschulschriften GmbH & Co. KG and licensors
All rights reserved. Saarbrücken 2012

Table of Content

1 Introduction 5
2 Fundamentals 9
2.1 Solid Liquid Interface and Electrochemical Double Layer 9
2.2 Fundamentals of Electrode Reactions 11
2.3 Electrochemical Methods 13
2.4 Scanning Probe Microscopy 17
    2.4.1 STM and EC-STM 17
    2.4.2 AFM 19
    2.4.3 SECPM 19
2.5 Single Crystals and Single Crystalline Supports 21
    2.5.1 Platinum 22
    2.5.2 Gold 24
    2.5.3 Palladium 25
    2.5.4 Rhodium 25
    2.5.5 Iridium 26
    2.5.6 Ruthenium 26
    2.5.7 Highly Oriented Pyrolytic Graphite (HOPG) 26
2.6 Deposition Techniques 27
    2.6.1 Underpotential and Overpotential Deposition of Metals 27
    2.6.2 Metal Nucleation and Growth 28
    2.6.3 Single Particle Deposition 29
2.7 Hydrogen Evolution / Hydrogen Oxidation Reaction (HER)/(HOR) ... 30
2.8 Oxygen Reduction Reaction (ORR) 33
2.9 Methanol Oxidation Reaction (MOR) 36
3 Materials and Methods 38
3.1 Chemicals 38
3.2 Electrochemical Setup 39
    3.2.1 Electrochemical Glass Cell 39
    3.2.2 Potentiostats 39
3.3 Electrochemical and SPM Techniques 39
    3.3.1 Setup EC-SPMs, (STM, AFM, SECPM) 39
    3.3.2 Tip Preparation 40
    3.3.3 Reference Electrodes for EC-SPM 43
    3.3.4 Local Reactivity Measurements 43
    3.3.5 Potential Behavior of Pd/H Electrodes – pH Dependency 45
3.4 Electrodes and Electrode Preparation 46
    3.4.1 Au(111) Single Crystal and Au(111) Films on Mica 46
    3.4.2 HOPG 47
    3.4.3 Ru(0001), Rh(111), Ir(111) and Ir(100) Single Crystalline Films 47
3.5 Deposition Techniques 47
    3.5.1 Single Particles with an STM 47
    3.5.2 Electrochemical Metal Deposition from Solution 48
4 Results 50
4.1 EC-SPM on Au(111), HOPG, Ru(0001), Rh(111), Ir(111), Ir(100) 50
    4.1.1 HOPG and Au(111) 50
    4.1.2 Ru(0001) 53
    4.1.3 Rh(111) 55
    4.1.4 Ir(111) and Ir(100) 57
4.2 Nanostructured Model Surfaces 59

    4.2.1 Pulse Deposition Techniques 59
    4.2.2 Pd on Au(111) 60
    4.2.3 Pt on Au(111) 61
    4.2.4 Pt on Ru(0001) 64
    4.2.5 Cu-upd in Perchloric and Sulfuric Acid on Rh(111) 66
4.3 HER/HOR, ORR and MOR on Pd/Au(111) and Pt/Au(111) 69
    4.3.1 HER/HOR on Pd/Au(111) 69
    4.3.2 HER/HOR on Pt/Au(111) 70
    4.3.3 ORR on Pd/Au(111) 71
    4.3.4 ORR on Pt/Au(111) 72
    4.3.5 MOR on Pd/Au(111) and Pt/Au(111) 73
4.4 Local pH Sensor, Pd/H Electrode 75
5 Discussion 80
5.1 New Supports in Electrochemical Environment 80
5.2 Local Approaches for Reactivity Measurements 81
    5.2.1 Local Current Technique 81
    5.2.2 Local Potential Technique 82
5.3 Support Effects 83
    5.3.1 Hydrogen Reactions – Specific Reactivity 83
    5.3.2 Geometric Current Density of HOR/HER of Pd and Pt on Au(111) 87
    5.3.3 Single Particles vs. Large Nanostructured Electrodes –Pt/Au(111) 89
    5.3.4 Oxygen Reduction Reaction 91
    5.3.5 Methanol Oxidation Reaction 92
    5.3.6 Summary of Support Effects 93
6 Summary and Conclusions 94
7 References 96
Appendix 104
A1 Abbreviations and Symbols 104
A2 Publications 106

1 Introduction

Energy production, conversion and storage due to changed energy demand are essential challenges today. All energy related fields have to improve and rationalize their techniques and methods. In the case of electrochemical energy storage and conversion the electrolysis and the fuel cell technology play important roles in the energy flux. Fuel cells convert chemical energy into electricity which is possible by separating the oxidation and reduction reactions when oxidizing a fuel. Electrolyzers convert electricity into chemical energy. Depending on the reaction pathway and the reaction partners usually a fairly high activation barrier is given. Catalysts lower these barriers and therefore promote the reactions. In general, noble metal catalysts such as platinum or platinum containing alloys are the best known catalysts especially for low temperature applications, e.g. polymer electrolyte fuel cells. In this context, the hydrogen evolution/hydrogen oxidation, oxygen reduction and methanol oxidation are under detailed investigation. Starting from a fundamental point of view model nanostructured surfaces were used to investigate their catalytic reactions towards the above mentioned reactions.

Several parameters determine the activity of the noble metal catalysts such as particle size [1-3], particle dispersion [4, 5], density of low coordinated surface atoms [6, 7] and influence of the substrate material [8-17]. Fast transport due to spherical diffusion for small, isolated catalyst particles resulting from a fine particle dispersion might also be an important parameter [5, 18]. One distinct property of nanometer sized particles is their large specific surface area and the large number of low coordinated sites on the surface, which may have a direct impact on their reactivity. For example, the variation of platinum nanoparticle size results for several reactions in considerable changes in the particle reactivity, e.g. specific activity towards the hydrogen evolution/hydrogen oxidation [4, 5, 18-21], oxidation of methanol [22, 23] and formic acid [23] and the reduction of oxygen [24, 25]. But it was also found that nanoparticles can not be described just by the sum of the behavior of the adsorption sites they provide determined by the coordination of the atoms. This is for example evident from infrared stretching frequency, from nuclear magnetic resonance studies of linearly adsorbed CO on Pt nanoparticles of different size and from density functional theory calculations [3, 26-29].

Nanoparticles can also change their electronic structure when supported on a foreign support and differ significantly from bulk material behavior. According to theoretical calculations of the group of Nørskov [9, 30-36], the group of Groß [11, 12, 37] and the group of Schmickler [38-42] the electronic structure of metals influences their reactivity when supported on or alloyed with other metals. Density Functional Theory (DFT) is the approach to calculate activation energies and adsorption energies of species and reaction intermediates on surfaces [43]. For example the electrocatalytic properties of Pd islands on Au(111) are modified due to strain in the lattice of Pd by about 4.8% resulting in a shift of the d-band centre of Pd to higher energies and thus a change in the interaction with reaction intermediates, e.g. with atomic hydrogen [21]. Similar conclusions were taken by Roudgar and Gross [11] who investigated theoretically the adsorption energies of hydrogen and CO.

It was experimentally shown by various groups [13, 17, 44-48] that the physical and chemical properties of thin Pd overlayers change with respect to the Pd bulk material. The reactivity for hydrogen reactions, oxygen reduction and formic acid oxidation is obviously dependent on the thickness of the Pd layer, the crystallographic orientation and also on the chemical identity of the substrate. Most of the observed experimental results can be explained by a lateral strain of thin Pd films on Au(111) electrodes according to the Nørskov model [35]. Pandelov and Stimming [16] more recently reported an increased specific reactivity of Pd submonolayers on Au(111) which is two orders of magnitude higher compared to bulk Pd. A spill-over mechanism of the adsorbed hydrogen during the HER from the Pd sites to the Au substrate was able to explain the results for submonolayer coverage [15]. Local reactivity measurements and combined DFT calculations of single Pd nanoparticles on Au electrode surfaces were performed by Meier et al. [19, 21] and show increasing activity with decreasing particle height.

For a closer look to the nano world and to get a direct or indirect access up to the atomic scale many techniques and instruments were invented over the past years. Since the invention of the Scanning Tunneling Microscopy (STM) by Binnig and Rohrer et al. [49, 50] atomic resolution of an electron conducting sample was possible. This milestone was awarded with the Nobel Prize in the year 1986. Later on the STM was transformed from the original ultra high vacuum technique to normal air mode by Haiss et al. [51] and afterwards based on the developments of bipotentiostates to electrochemical conditions (see [52, 53]). Further developments in the scanning probe microscopy were established, Atomic Force Microscopy (AFM), Magnetic Force Microscopy (MFM) and Near Field Scanning Optical Microscopy (SNOM) are some of the commonly used techniques [54-57]. With all these concepts a direct acquisition and controlled preparation in the nanometer scale at single particles is possible.

All above mentioned work contributes to a better understanding of the parameters influencing the electrocatalytic properties of catalysts, which is important for a rational design of catalysts. Various parameters such as the interparticle distance, particle morphology, chemical composition and influence of the support influence the behavior of the catalyst. In an effort to better separate the various parameters; this thesis follows the approach of depositing nanoislands of different reactive materials such as Pd and Pt onto nonreactive model surfaces, for instance Au(111) and Ru(0001). The electrocatalytic behavior of Pt and Pd nanoparticles on Au(111) systems concerning the Hydrogen Oxidation Reaction/Hydrogen Evolution Reaction (HOR)/(HER), the Oxygen Reduction Reaction (ORR) and the Methanol Oxidation Reaction (MOR) were investigated in detail. First results on Pt/Ru(0001) and Cu-upd on Rh(111) will be shown. Electrochemical methods (e.g. potentiostatic pulse techniques and cyclic voltammetry) are used in order to study the influence of the above mentioned parameters on HOR/HER, ORR and MOR. With *in-situ* electrochemical STM (EC-STM) morphological and structural properties of the model catalysts were investigated in parallel to reactivity measurements. New supports for electrochemical investigations were characterized with standard techniques and a new developed method called scanning electrochemical potential microscopy (SECPM). Also a local pH sensor for investigations in the nm range will be presented. This combined approach will help to obtain a more detailed picture of structural and morphological parameters influencing reactivity.

After the introduction in Chapter 1, Chapter 2 introduces fundamentals of electrochemistry. Applied electrochemical as well as scanning probe methods and the different supports are introduced. A detailed review of the investigated reactions is also given. Especially the electrochemical double layer will be introduced to explain later also the working principle of the SECPM technique. Principles of reactions occurring at an electrode in an electrochemical environment with corresponding treatment of mathematical formalism regarding the relationship of potential and current are given. Also the used measurement techniques such as cyclic voltammetry and pulse methods are presented. EC-STM, SECPM and AFM will be introduced and the applications in electrochemistry are shown. The Chapter will be completed by reviews of the hydrogen, oxygen and methanol reaction whereas the hydrogen reaction will play a major role.

Chapter 3 includes information about chemicals, experimental setups and conditions. The used electrochemical as well as scanning probe equipment is introduced. Experimental procedures such local reactivity measurements and potential behavior of palladium hydrogen electrodes are described. Electrode preparation as well as electrochemical metal deposition will be specified.

Hydrogen reactions on Pd and Pt decorated Au(111) single crystal surfaces will be the first part of the results in Chapter 4, followed by a detailed presentation of results regarding the oxygen and methanol reaction on the same surfaces. SPM and electrochemical studies on different new single crystalline supports such as Ru, Rh and Ir for electrochemical investigations are shown. The behavior of a local pH sensitive electrode in form of an insulated hydrogen loaded Pd tip is also presented for purposes in the SECPM.

In the discussion in Chapter 5 effects for supported noble metal particles such Pd and Pt overlayers on Au(111) will be compared to bare noble metal surfaces. Due to the different characteristics of hydrogen and oxygen/methanol reactions observed on the investigated model surfaces different models and explanations will be given and applied as far as possible. Especially for the hydrogen reaction a detailed comparison of Pt single particle measurements with large extended Pt decorated Au(111) will be given. Details about the experimental techniques will help to get a closer look insight the fundamentals of hydrogen electrocatalysis. The SECPM as promising new technique will be discussed in terms of monitoring different support materials for model systems and preliminary results on a local potential shift method will be shown. A short section about the investigated new single crystalline support materials for fundamental investigations in electrocatalysis and studies concerning adsorption processes will complete the discussion section.

Chapter 6 summarizes the results and the discussion. Some conclusions follow and an outlook will be given. Basic concepts of the work and achieved results are highlighted. Statements which can be extracted from the findings in this work are used to depict further research in the field of fundamental aspects of substrate effects in electrocatalysis.

2 Fundamentals

2.1 Solid Liquid Interface and Electrochemical Double Layer

Electrochemical investigations can be defined as "the study of structures and processes at the interface between an electronic conductor (the electrode) and an ionic conductor (the electrolyte)" [58]. The structure of the interface and the reactions that occur are of great interest for e.g. electrochemical energy conversion and energy storage. Electrochemical methods can determine potentials, currents, charges and/or capacitances to have access to fundamental processes. In this section, principles of electrochemistry, the structure of the solid/liquid interface and electron transfer processes will be introduced.

When a metal surface is immersed in a liquid the interfacial double layer is formed due to compensation reactions until the equilibrium between these two phases is achieved. This equilibrium on the solid/liquid interface is associated to a potential difference between the electrode and the electrolyte. Excess charge on the metal electrode is then compensated by ionic charge in the electrolyte. In 1853, Helmholtz [59] described the solid/liquid interface as a rigid double layer comparable to a parallel plate capacitor considering the electrode (metal) surface as one plate and the electrolyte as the other plate holding the counter-charge. This simple model of two layers of opposite charge was the origin of the term "electrochemical double layer" (EDL). According to Helmholtz, the double layer has a differential capacitance C_H which is given by

$$C_H = \frac{\varepsilon \varepsilon_0}{d} \qquad (2.1)$$

where $\varepsilon > 1$ is the dielectric constant modulating the permittivity of free space ε_0 and d is the distance between the two virtual plates. Typical values of C_H lie between 20 to 40 $\mu F/cm^2$, which correspond to experimental values. The constant capacitance in the Helmholtz Model is an early approximation of the solid-liquid interface which does not completely describe real systems. Here, the experimental results vary with the applied potential or the concentration of the electrolyte.

Gouy [60] and Chapman [61] independently developed theories including the thermal motion of ions in the EDL in the beginning of the 1920s. A diffuse layer within the electrolyte was proposed which can be approximated by a series of laminae parallel to the electrode. The electrostatic potential varies with distance from the electrode and thus influences the energy of the ions in different laminae. The ion species are treated as point particles and thus follow a Boltzmann distribution. Combining this assumption with the Poisson equation yields the Poisson-Boltzmann equation which is valid when the dielectric constant of the solvent is independent of distance and ion concentration. For a symmetrical electrolyte the Poisson-Boltzmann equation [62, 63] can be simplified to:

$$\frac{\tanh(\frac{ze\varphi(x)}{4k_BT})}{\tanh(\frac{ze\varphi_0}{4k_BT})} = \exp(-\kappa x) \quad \text{with} \quad \kappa = (\frac{2n^0z^2e^2}{\varepsilon\varepsilon_0 k_B T})^{1/2} \quad (2.2)$$

κ is the reciprocal of the characteristic diffuse double layer thickness, called Debye-Hückel. (z : charge magnitude, e : charge of an electron, k_B : Boltzmann constant, T : Temperature, φ : potential, x : distance from electrode, n : concentration). The electrochemical double layer can be approximated with equation $\tanh(\xi) \approx \xi$ for small ξ by:

$$\varphi(x) = \varphi_0 \exp(-\kappa x) \quad (2.3)$$

The electrode surface charge density, the charge density in the solution phase at the interfaces, the concentration profile of each ion species and the differential capacitance C_d can be obtained by calculation of φ from equation (2.2) or (2.3) and thermodynamic formalism.

While Gouy and Chapman considered the ions as point charges Stern improved the model by taking the finite size of the ions into account, i.e. the ions can not approach the electrode surface closer than their added ionic radius. If they are solvated, the radius of the solvation shell is limiting their distance. This distance of closest approach for the centers of the ions is x_2 which defines the outer Helmholtz plane (OHP), where φ_2 is the potential in the OHP. The ions closest to the electrode forming the OHP and that being held in position by purely electrostatic forces are termed "nonspecifically adsorbed" ions. These are mainly solvated cations. At distance $x > x_2$ the considerations of the Gouy-Chapman theory, especially the Poisson-Boltzmann equation remain valid. Stern's interfacial model can be treated as an extension for $x \leq x_2$ where no charge between the electrode surface (x_0, φ_0) and the OHP (x_2, φ_2) is present. Thus, according to the Poisson equation the potential drops linearly from φ_0 to φ_2. The potential profile in the diffuse layer of a symmetrical electrolyte is similar to equation (2.3) whereas $(-\kappa x)$ is replaced with $(-\kappa(x-x_2))$. The differential capacitance C_d can be obtained from the thermodynamics formalism (for detailed information see [53]).

$$\frac{1}{C_d} = \frac{x_2}{\varepsilon\varepsilon_0} + \frac{1}{\varepsilon\varepsilon_0 \kappa \cosh(\frac{ze\varphi_2}{2k_BT})} = \frac{1}{C_H} + \frac{1}{C_D} \quad (2.4)$$

This model is known as the Gouy-Chapman-Stern (GCS) model schematically shown in Figure 2.1.

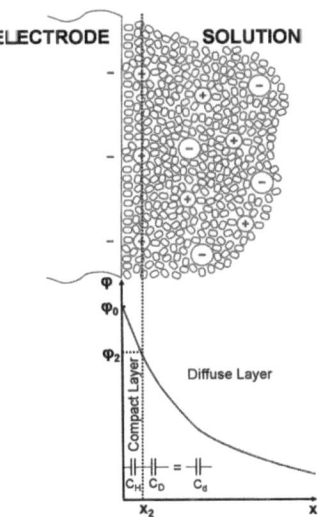

Figure 2. 1 Scheme of the composition and the potential profile of the electrochemical double layer according to the GCS model, based on [64] (upper part) and [62] (lower part).

As it can be clearly seen in equation (2.4) the double layer capacitance consists of two contributions: C_H and C_D. C_H is the contribution of the rigid Helmholtz layer which does not depend on the potential whereas C_D is the contribution of the diffuse layer.

Although being a good approximation for the electrochemical double layer the Gouy - Chapman - Stern theory only describes electrostatic effects while disregarding polarization effects within the electrolyte, i.e. the dependence of the dielectric constant on the distance to the electrode or the dependence of C_H on the potential. Furthermore, the model neglects ion-ion interactions or strong nonspecific interactions of the ions with the electrode surface. Usually the capacitance of the double layer can not be neglected during fast electrochemical experiments when a charging current of electrochemical double layer is flowing and overlapping the current signal. Hence, the current for the charging process has to be recalculated to separate the Faradaic current which is due to electrochemical reactions from the double layer current [62, 63].

2.2 Fundamentals of Electrode Reactions

An electrode reaction is a process that involves the transfer of electrons from an electrode to a chemical species, or vice versa. In the following, an electrode is immersed in an electrolyte that contains the electroactive species I with its two different oxidation states, oxidized Ox and reduced

Red, where n is the number of the transferred electrons within the reaction. A simple electrode reaction that converts Ox to Red and vice versa can be described by

$$Ox + ne^- \leftrightarrow Red \qquad (2.5)$$

and consists of three steps: 1) the reagent Ox must reach the electrode surface, 2) the heterogeneous electron transfer process from the electrode to the species Ox takes place and 3) the reaction product Red must leave the electrode surface. The overall reaction rate will be limited by the slowest elementary step either caused by mass transport (step 1 and step 3) or electron transfer (step 2). Electrode reactions can involve multiple electron transfers e.g. metal deposition, corrosion or coupled chemical reactions.

Any combination of two electrodes connected via an ionic conductor is called galvanic cell. Without any current flowing, the voltage built up between these two electrodes is called open circuit voltage (OCV). The Nernst equation describes the dependence of the open circuit voltage on the concentration of the electrolyte and is therefore of great importance in electrochemistry. Based on the difference of the electrochemical potential $\mu_i = zF\Delta\varphi$ between the electrodes and the electrolyte, the Nernst equation can be expressed as

$$U_0 = U_{00} + \frac{RT}{nF} \ln \prod_{i=1}^{m} [a_i]^{v_i} \qquad (2.6)$$

or concentration dependent as

$$U_0 = U_{00} + \frac{RT}{nF} \ln \frac{c_{ox}}{c_{red}} \qquad (2.7)$$

where U_{00} is the standard potential, R the universal gas constant, T the temperature, n the number of transferred electrons, a_i the standard activity of species i, v_i the stochiometric value of species i in the reaction equation and c_{ox}/c_{red} the concentration of the oxidized/reduced species.

Regarding the pH dependency measurements of the Pd/H electrode in this thesis, the Nernst equation is one of the basic equations to calculate the voltage variation while changing the pH value of the solution. The Nernst equation describes equilibrium states with no current flow. Applying an overpotential η causes a disturbance of the equilibrium which results in a current flow.

The correlation between the applied electrode potential U and the current density j, which is based only on thermodynamics, is described by the Butler-Volmer equation

$$j(\eta) = j_0 \left[\exp\left(\frac{\alpha_a nF}{RT}\eta\right) - \exp\left(-\frac{\alpha_c nF}{RT}\eta\right) \right] \qquad (2.8)$$

Where the overpotential η is defined as $\eta = U - U_0$, i.e. the deviation of the potential from the equilibrium potential, α_a and α_c are the transfer coefficients for the anodic and the cathodic reactions, respectively. The Butler-Volmer equation is the fundamental equation of electrode kinetics and can be reduced for three limit cases, whereas j_a is the anodic current and j_c is the cathodic current:

- high positive overpotential: $|j_a| \gg |j_c|$

$$\log j = \log j_0 + \frac{\alpha_a nF}{2.3RT}\eta \qquad (2.9)$$

- high negative overpotential: $|j_c| \gg |j_a|$

$$\log- = \log j_0 - \frac{\alpha_c nF}{2.3RT}\eta \qquad (2.10)$$

- very low overpotential: $\eta \ll (RT)/(\alpha_a nF)$ and $\eta \ll (RT)/(\alpha_c nF)$

$$j = j_0 \frac{nF}{RT}\eta \qquad (2.11)$$

Equations (2.9) and (2.10) are known as Tafel equations and are used to determine the exchange current density j_0 and the transfer coefficients α_a/α_c. The linear approximation in equation (2.11) will be used to determine the exchange current density using small overpotentials near the equilibrium, so called micropolarisation curves.

2.3 Electrochemical Methods

Potential sweep techniques, such as linear sweep voltammetry (LSV) and especially cyclic voltammetry (CV), have become very popular techniques for initial electrochemical studies of new systems and thus, have been applied to an ever increasing range of systems [62, 65]. With both techniques the cell current I is recorded as a function of the applied potential U. Usually the excitation signal is a potential ramp between different potentials U_i, i.e. the potential is linearly varied with time at known sweep rates v. While LSV sweeps the electrode potential from potential U_1 to U_2, CV inverts the direction when reaching the potential U_2 and the potential sweeps back to U_1. The potential sweep can be stopped or again swept to U_2. The electrochemical data obtained in such experiments yields information about processes occurring at applied potentials. A typical triangular waveform of the excitation signal in a CV experiment is shown in [62, 63] as well as the response of an electrochemical system with an electroactive species Ox in solution that can be reduced to Red. Since the mathematical description of these techniques is well developed, kinetic parameters for slow reactions can be determined for mechanisms of fairly complicated electrode reactions. From the sweep rate dependence the involvement of coupled homogeneous reactions can be identified and processes such as adsorption can be recognized. The analysis of kinetics for fast reactions of electrochemical reactions at the electrodes requires a separation of the kinetic effects from the inhibited transportation elements which can not be easily obtained with CVs. Thereby potential step and current step methods were applied to have access to reactions with fast kinetics or special requirements for deposition and adsorption or absorption processes.

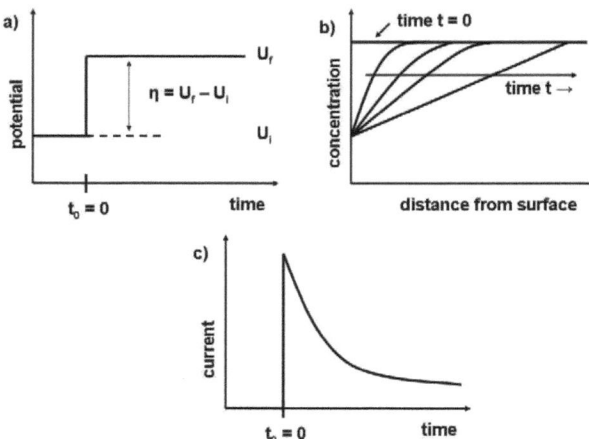

Figure 2. 2 Schematic potential pulse step; (a) potential pulse versus time; (b) growth of diffusion layer; (c) current transient.

Potential step experiments cause current-time transients which were recorded and extrapolated to the initial time t = 0 where idealized no diffusion effects occur. Double layer charging in the first few micro seconds can also be considered during fitting. At this first step of electrochemical reaction the concentration profile is very steep and the concentration profile is not formed (see Figure 2.2 b). Hence substance transportation is fast and the reaction rate is determined by the kinetics. For this technique rapid measurement and data acquisition methods are necessary. The basic equations are given in the following:

A potential pulse at an electrode is schematically shown in Figure 2.2 a). At time t = 0 the potential jumps from the initial potential U_i to the final potential U_f and a constant concentration on the surface is forced. Figure 2.2 c) shows the current transient due to the potential step. The current is a result of three different processes: Double layer charging, kinetic current and diffusion limitation. If chemical reactions, adsorption or desorption processes and multistep reactions are neglected kinetic and diffusion processes can be precisely investigated. With this basic approach the diffusion equation (Fick's law) for both species (see ref [63]):

$$\frac{\partial c_{ox}}{\partial t} = D_{ox} \frac{\partial^2 c_{ox}}{\partial x^2} \quad \text{and} \quad \frac{\partial c_{red}}{\partial t} = D_{red} \frac{\partial^2 c_{red}}{\partial x^2} \qquad (2.12)$$

in connection with the concentration dependent Butler Volmer equation

$$j = j_0 \left\{ \frac{c_{red}^s}{c_{ox}^s} \exp\left[\frac{\alpha_a \, n \, F}{R \, T} \eta\right] - \frac{c_{ox}^s}{c_{red}^s} \exp\left[-\frac{\alpha_c \, n \, F}{R \, T} \eta\right] \right\} \quad (2.13)$$

and special boundary conditions will give the following solution for the current density:

$$j = j_0 \left\{ \exp\left[\frac{\alpha_a \, n \, F}{R \, T} \eta\right] - \exp\left[-\frac{\alpha_c \, n \, F}{R \, T} \eta\right] \right\} \exp\left[\lambda^2 t\right] \operatorname{erfc}\left[\lambda \sqrt{t}\right] \quad (2.14)$$

whereas

$$\lambda = \frac{j_0}{n \, F} \left\{ \frac{\exp[\alpha_a \, n \, F \, \eta / R \, T]}{c_{red}^0 \sqrt{D_{red}}} + \frac{\exp[-\alpha_c \, n \, F \, \eta / R \, T]}{c_{ox}^0 \sqrt{D_{ox}}} \right\}. \quad (2.15)$$

The solution can be divided into two parts, one time independent part and one time and overpotential dependent part. The time independent part represents the kinetic current density j_D because for $t \to 0$ the time and overpotential dependent part results to be zero, $\lim \operatorname{erfc}(\lambda \sqrt{t}) = 1$ for $t \to 0$ (erfc = error function). Hence at the time $t = 0$ one only derives the kinetic current j_D. For long times the time dependent part can be approximated with $1/\lambda(\pi t)^{1/2}$. The diffusion current is therefore proportional to $1/t^{1/2}$. In contrast to that, the kinetic current results from an extrapolation from the current time curve to the time $t = 0$. With a short time approximation $\lambda(t)^{1/2} \ll 1$ one gets an estimation for $\exp(\lambda^2 t) \operatorname{erfc}(\lambda(t)^{1/2}) = 1 - 2\lambda(t/\lambda)^{1/2}$ which results in the following expression:

$$j(\eta, t) = j_0 \left\{ \exp\left[\frac{\alpha_a \, n \, F \, \eta}{R \, T}\right] - \exp\left[\frac{-\alpha_c \, n \, F \, \eta}{R \, T}\right] \right\} \left(1 - 2\lambda\sqrt{t/\pi}\right)$$

$$= j_D - 2j_D \frac{\lambda}{\sqrt{\pi}} \sqrt{t}. \quad (2.16)$$

For an evaluation process the current density is plotted versus $\sqrt{t}$. For times below 1ms the analysis gets more difficult because the charging of the double layer has to be considered.. A typical current transient for the kinetic investigation for the HOR is presented in Figure 2.4 (I). A potentiostatic pulse with a height of an overpotential of +130 mV was applied to a Pt(111) crystal in 1 M HClO$_4$. In order to extract the kinetic current the current transient is plotted versus the square root of the time as shown in Figure 2.3 (II) [62, 63]. Extrapolation from the first milliseconds after the potential step to $t = 0$ yields the kinetic current density (j_{kin} = 0.066Acm^{-2}), which is shown in Figure 2.3 (II). With this method it is possible to evaluate kinetic current densities j_0 in the range less than 50mA/cm^2. The kinetic currents obtained are referred to the geometric electrode surface but were also related to the surface of deposited Pd and Pt as specific current densities in the case of Pd/Au(111) and Pt/Au(111) nanostructured surfaces. Figure 2.3 (III) shows the current density vs.

one over square root of time in order to obtain the diffusion limited current as intercept of the transient with the y-axis. Especially for high current densities the correction of the IR drop, i.e. the potential drop caused by the resistance of the electrolyte can not be neglected [63]. The electrochemical double layer charging as a fast process occurring during the first 100µs of the measurement can be separated from the transient during evaluation.

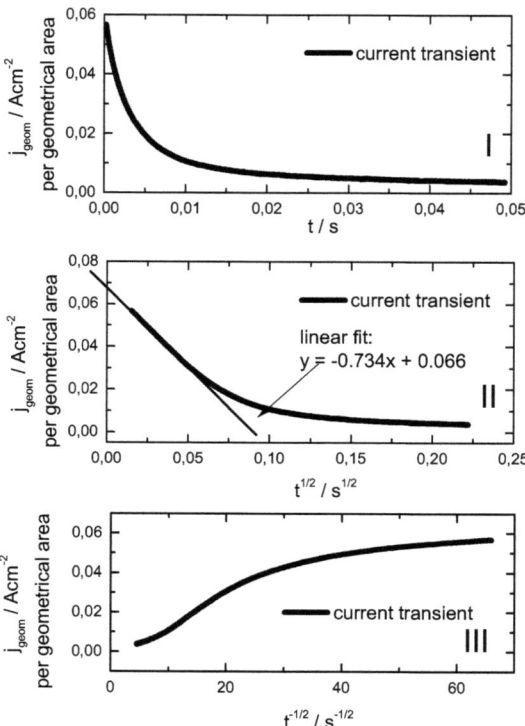

Figure 2. 3 Series of the evaluation of potentiostatic pulse method with current density vs. time (I), current density vs. square root of time (II) and current density vs. one over square root of time (III) for a Pt(111) single crystals in 1M HClO₄. The applied potential was +130mV vs. the hydrogen equilibrium.

During the current step method a current pulse is applied at the electrode und thus inducing a potential response (see Figure 2.4). This technique is comparable to the potential step method whereas the driving force is a potential pulse and a current transient is recorded. A galvanostat is used to control the applied current on the system. Diffusion and charge transfer controlled systems can be investigated and the processes can be separated.

The current is changed from I_i to I_f inducing a change of the electrode potential. In an idealized case the potential jumps to a value where a reaction occurs at the electrode consuming the preset current. The potential increases if the concentration of the reacting species decreases or if the surfaces reaction needs higher potential to occur with the same applied current, shown in Figure 2.4 with a solid line. A real gradient is shown with the dotted line where the electrochemical double layer has to be charged and thus the potential increases not as abrupt as in the idealized case. This observation is comparable to the one for the potential pulse where the double layer charging has also to be considered in the current transient. All given formula can also be applied to that technique by replacing the potential with the current in the different steps. More details are given in [62, 63].

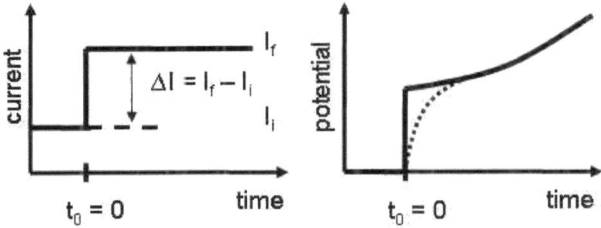

Figure 2. 4 Schematic current pulse step: (A) current vs. time; (B) potential vs. time.

2.4 Scanning Probe Microscopy

2.4.1 STM and EC-STM

The invention of the scanning tunneling microscopy (STM) by Binnig et al.[49, 50] in 1982 paved the way for a variety of scanning probe techniques and the investigation of surfaces on a nanometer scale [66-72]. Since Hansma and co-workers [52, 53] demonstrated that STM can operate in electrolyte solutions, much progress has been made to explore the potential of the electrochemical STM (EC-STM). Especially, the development of a four electrode configuration in the STM in order to carry out complete *In-Situ* experiments under potentiostatic control [73, 74] has enabled new perspectives for studying electrochemical processes at the solid-liquid interface on a nanometer or even atomic scale.

The tunneling current that serves as feedback signal in the STM setup is attributed to the quantum mechanical tunneling effect which occurs when the wave functions of a metallic tip and the substrate atoms overlap, *i.e.*, the STM maps the local density of states (LDOS). Thereby, electrons are able to tunnel through a potential barrier consisting of extremely thin, insulating layers such as vacuum, air or liquids with a potential barrier height between 1 eV and 4 eV [75, 76] The tunneling current is strongly dependent on the thickness of this layer = distance tip – sample. STM images are recorded while the tip scans in x-y direction over the sample surface and the tunneling current flows perpendicular to the surface. Depending on the operation mode a feedback loop controls the movement of the tip in order to maintain a constant tunneling current (constant current mode) or the tip scans without feedback control, keeping the distance between tip and substrate constant, and measuring the tunneling current (constant height mode). In general STM images show a convolution of electronic properties and the topography of the sample surface.

Figure 2.6 shows the experimental setup of the electrochemical scanning tunneling microscope. Following the typical arrangement in an electrochemical cell with a working electrode (WE), counter electrode (CE) and reference electrode (RE) are added to measure the current and adjust the potential. The potentiostat allows a defined control of potential between WE and RE due to high input impedance whereas the current is supplied via the CE. Due to the limited space in the EC-STM cell both electrodes are metallic wires typically made of Pt, Pt/Ir, Pd or Au. In addition to the sample electrode, the tip in contact with the electrolyte acts as a fourth electrode which has also to be potential controlled in order to precisely apply the bias voltage. A bipotentiostat is able to adjust the potentials at the tip U_{tip} and the sample U_s independently versus a reference electrode. The difference $U_{tip} - U_s$ is defined as U_{bias} [145, 151]. While the bipotentiostat is adjusting the electrode potentials and thus for the electrochemical processes in the STM cell, the control processing unit (CPU) of the STM controls the movement of the tip, and detects the measured tunneling current. Due to the fact that both working electrodes, the sample and the tip are in contact with the electrolyte Faradaic currents may occur at both surfaces. While small Faradaic currents at the sample surface I_s do not disturb the measurement, a Faradaic current at the tip can easily overlap the tunneling current and make the STM measurement impossible. Since tunneling currents are normally adjusted between 0.2 nA and 1 nA the Faradaic leakage currents should not exceed 0.1 nA. In order to reduce the Faradaic currents to a few pA the metal tip is insulated with Apiezon wax to minimize the free surface area. Furthermore, STM tips are usually made of metals that are electrochemically stable over a large potential range such as gold, palladium and platinum. However, the measured current at the tip I_{tip} is always a sum of the tunneling current and Faradaic currents.

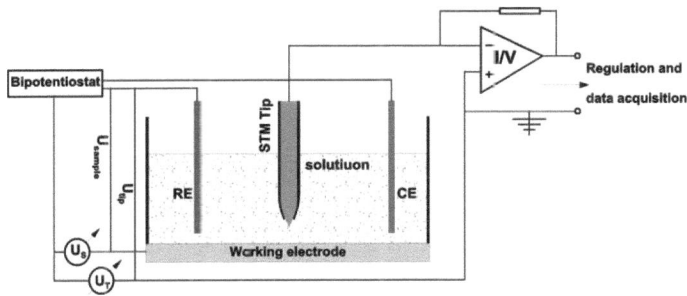

Figure 2. 5 Principles of an electrochemical STM from [77].

2.4.2 AFM

In 1986 Binnig et al. [78] introduced the atomic force microscope (AFM). The AFM probe is a cantilever with a sharp tip normally made of silicon or silicon nitride. While scanning the surface, AFM measures interaction forces between the tip and the sample surface. Since the nature of these forces can be electrostatic, van der Waals, frictional, capillary or magnetic, AFM is also suitable for the characterization of non-conductive surfaces and biological samples. The repulsive and attractive interaction forces, usually in the range between nN and µN, cause the deflection of the cantilever which is typically measured using a laser [79]. Thereby, the laser beam is focused on top of the cantilever and reflected into an array of photodiodes resulting in an electrical signal that is fed into the feedback circuit. When the deflection of the cantilever changes the reflected spot moves on the photodiode and changes the electrical signal.

AFM imaging modes can be divided into static and dynamic modes. Being operated in the static mode, the AFM tip touches the sample and senses repulsive forces (contact mode). In the dynamic mode, the cantilever oscillates preventing the tip from touching the surface and the AFM tip senses attractive forces (non-contact mode). This mode avoids that the tip or the surface of the sample are damaged which results in a decreased lateral resolution compared to contact mode. The attractive forces result in an amplitude, frequency or phase shift of the tip oscillation, which can be used as feedback control parameter to map the surface [80-82].

2.4.3 SECPM

In 2004 Woo et al. [83] reported on a modified EC-STM with a miniaturized potential probe in order to measure the local potential of solid-liquid interfaces with subnanometer spatial resolution. In 2007 this technique, now termed scanning electrochemical potential microscopy (SECPM), was established in the group of Allen J. Bard [84], where STM was also used to a large extent to investigate supports such as Au [85], HOPG [86] and Cu [87].

In order to determine the electrode surface potential in polar liquids or electrolytes, SECPM uses the potential gradient present in the electrochemical double layer (EDL) formed at the solid-liquid interface. The hardware is similar to an EC-STM; the only modification consists of replacing the current pre-amplifier by a high input impedance potential difference amplifier. Since both imaging techniques STM and SECPM are implemented in one head combined EC-STM/SECPM, studies of the same area of an electrode are possible, *i.e.,* potential maps of the surface obtained at constant potential SECPM can be directly compared to the images of the local density of states (LDOS) obtained in constant current mode of the electrochemical STM.

Recently, it was shown that SECPM can be utilized to investigate the metal content and distribution of tungsten modified diamond-like carbon (DLC) films [88]. Furthermore, SECPM is a promising technique for imaging biological samples such as enzymes and proteins immobilized on electrode surfaces providing a resolution higher than EC-STM [89]. Since no electron transfer between the low conductive biomolecule and the electrode is necessary in the case of SECPM, potential mapping of horseradish peroxidase (HRP) adsorbed on HOPG, allowed identification of the heme group within the protein pocket.

In addition, SECPM also offers the possibility to map the potential distribution of the interface in x-z direction using the SECPM tip as potential probe [83, 84]. However, in order to interpret these potential curves quantitatively, it is necessary to understand how the presence of a probe influences the original EDL at the electrode or *vice versa*. Since the metallic potential probe in contact with the electrolyte also forms an EDL it can interact and overlap with the EDL at the electrode at close distances. The measured interfacial potential results from the overlap of both EDLs. Diffuse double layer interactions between two parallel plates [90], heterogeneously charged colloidal particles [91], or charged particles near surfaces [92], have already been considered in the past as a comprehensive theory for potential profiling. First attempts at an electrostatic approach, directly aimed at the SECPM experiments, using finite element method (FEM) simulation to compute the EDL potential, measured with the metallic probe, were reported by Hamou et al. [93, 94]. Double layer effects caused by the geometry of the tip apex and the free area of the tip are included in this model. It was found that the shape of the metallic apex affects the ion distribution in the nanogap resulting in an electroneutral region between tip apex and electrode. Consequently, an overall explanation for the SECPM technique cannot easily be given here. Further experimental, as well as theoretical approaches, have to be done to clarify the feasibilities and limitations of this new technique including all above mentioned aspects.

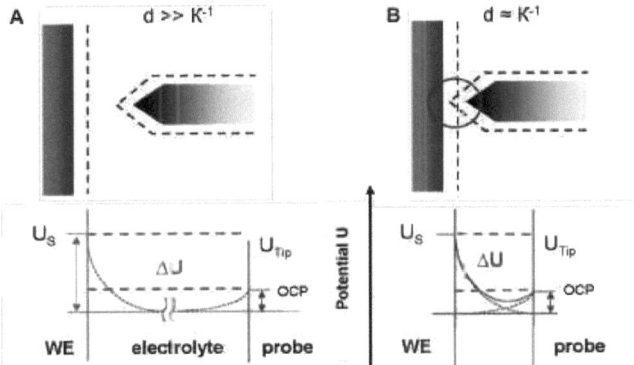

Figure 2. 6 Scheme of the EDLs present at the tip and the electrode surface, based on [83]: A) When tip and electrode surface are far apart the two EDLs do not influence each other. B) When the tip approaches the surface suddenly the EDLs starts overlapping [95].

In addition, SECPM also offers the possibility to map the potential distribution of the interface in x-z direction. According to the GCS model the theoretical potential of an electrode/electrolyte interface changes with the distance across the electrochemical double layer (EDL) [62]. Experimentally the SECPM tip serves as potential probe measuring the potential profile of an electrode while moving perpendicular to its surface (Figure 2.6). Thereby, the detected signal is given by the potential difference between the applied substrate potential U_s and the tip potential U_{tip}. If tip and electrode are in contact the potential measured at the tip is equal to the applied working electrode potential U. Far away from the surface, the potential of the EDL is equal to the potential of the bulk electrolyte. The decay of the EDL is characterized by the Debye length κ^{-1}.

2.5 Single Crystals and Single Crystalline Supports

Single crystal surfaces play an important role in surface science by providing well defined systems. Especially in the case of electrochemistry, usage of single crystals has improved the understanding of surface processes. These can be absorption and adsorption processes, oxidation or reduction of different molecules and surface changes such as reconstruction phenomena. Also fundamental reactions such as hydrogen and oxygen reactions as well as oxidation of small molecules were intensively studied during the last decades on polycrystalline as well as single crystal electrodes. Although the quality of single crystals [96-100] and experimental conditions rapidly increased in the 90s, fundamental studies started long before [101-105]. Because of the high catalytic activity noble metal surfaces such as Pt, Pd, Ir and Rh are of special interest. The main part of this section will be on single crystals, properties of different metals and the dependency of crystallographic orientation on electrocatalytic activity. Also Au, Ru and HOPG will be reviewed as possible support material. Due to the inertness and low activity, Au, Ru and HOPG surfaces are often used as support for depositing foreign species such as metals, enzymes, etc... For the use in electrochemistry there are several ways to prepare single crystals ranging from ultra high vacuum (UHV) preparation, electrochemical treatment, mechanical cutting and polishing or a combination of these [105-108]. Although in literature much work was done on single crystals, the preparation of single crystals is a challenging task. Therefore only Pt and Au single crystals with (111) orientation was used in this thesis. Pd was deposited onto Au(111) and therefore available from submonolayers to multilayers with the crystal orientation of the support. Ru, Rh and Ir surfaces with different orientations were used as epitaxial grown thin films on Si wafers.

This idea was based on the alternative route for the fabrication of single crystal metal surfaces by depositing thin, heteroepitaxial films on foreign substrates such as MgO [109] or Al_2O_3 [110]. Recently the used single crystal metal films of iridium, rhodium, platinum and ruthenium were successfully grown on Si(001) or Si(111) substrates using yttria-stabilized zirconia (YSZ) buffer layers [111]. Ir(001) surfaces provided the base for the first single crystal diamond films on Si(100) [112]. Another approach was followed with Au(111) films deposited on glass supports commercially available from Schroer GmbH. Annealing of the films results in (111) single crystal grains with random azimuthal orientation. Such high quality single crystalline surfaces were used for a wide range of investigations with different SPM and standard electrochemical methods. Results from literature describing the behavior of single crystals will be used to determine and evaluate the quality of the surfaces and the purposes for use in electrochemical investigations.

2.5.1 Platinum

Due to a high catalytic activity of Pt towards several reactions such as hydrogen evolution, hydrogen oxidation, oxygen reduction and methanol oxidation, Pt is used as catalyst in various applications. Studies on Pt started several decades ago on polycrystalline and single crystal surfaces. Annealing as preparation method for Pt single crystals is widely found in literature [106, 113]. It was also found that the cooling procedure after annealing process is important for a well-defined structure. By applying the right annealing and cooling procedures specific peaks in the hydrogen region in cyclic voltammograms [114-117] and typical surface structures can be observed via STM [118-120]. Pt polycrystalline electrodes as well as defined single crystals were electrochemically investigated regarding orientation and electrocatalytic activity during the last decades [101-105, 119, 121-129].

Voltammograms of Pt(100) and Pt(110) in 0.1 M H_2SO_4 with their specific characteristics can be found in [105, 119-122, 125-127]. Due to the importance of the Pt(111) surface some details will be given. The characteristic current peak at 0.43 V versus NHE in the voltammogram of Pt(111) in 0.1 M H_2SO_4 is caused by a rearrangement in the adsorbed sulfate ion adlayer [128] and can be seen as a parameter for the quality of the single crystal. The ordered (3x7) R19.1° adlayer of these sulfate ions can be observed via STM which was first demonstrated by Funtikov et al. [123, 129] and Braunschweig and Daum [130]. Comparing Pt(111) in 0.1 M $HClO_4$ and in 0.1 M H_2SO_4 shows there are no differences in cyclic voltammograms below 0.34 V versus NHE indicating a weak influence of anions in this region.

Pt surfaces were also studied regarding electrocatalytic activity on polycrystalline electrodes or not well defined single crystals [101-104]. The hydrogen reactions were intensively investigated on polycrystalline as well as single crystal surfaces. In early measurements no dependency on crystallographic orientation for electrocatalytic activity was found which was mainly attributed to poor quality single crystals and unfavorable experimental conditions. These findings were disproved by Markovic et al. [96, 97] and Barber et al. [98, 100]. They found that the reactivity of different Pt(hkl) surfaces towards hydrogen reactions is strongly influenced by surface orientation. It was clearly shown that in alkaline as well as acidic electrolytes the reactivity increases in the following order (111) < (100) < (110). In acidic solution the exchange current density of Pt(110) surfaces is 3 times higher compared to Pt(111). These findings are in line with the activation energies obtained from Arrhenius plots [96] showing the largest value of 18 kJ/mol for Pt(111) and the smallest value of 9.5 kJ/mol for the Pt(110) single crystal surface.

Arrhenius plots were also used to evaluate the hydrogen evolution in frozen aqueous electrolytes on different polycrystalline metal electrodes [131]. Frese et al. [131, 132] investigated the HER on polycrystalline Pt electrode in the liquid and in the frozen electrolyte in the temperature range between 293K and 176K. In liquid as well as in frozen phase the Volmer-Tafel mechanism occurs with the Tafel reaction as rate determining reaction. Extracting an Arrhenius plot from [131] a linear behavior is shown before and after the melting point of the electrolyte at around 227K. The activation energies are in the range of 15 kJ/mol for the liquid electrolyte and in the range of 27 kJ/mol in the frozen electrolyte, clearly seen in the change of the slope at the melting point. The value obtained in the liquid electrolyte for the investigated polycrystalline electrode is comparable to the average of the values obtained on single crystals seen in [96, 133].

Markovic et al. [134] followed an approach by investigating the hydrogen reactions at temperatures between 274K and 333K to obtain kinetic rates at different reaction conditions using a rotating disc electrode setup. In this approach the exchange current density was determined using micro polarization curves applying only small overpotentials in the range of +/- 10mV versus the equilibrium potential, a value of j_0 of approx. $1 mA/cm^{-2}$ was obtained. In addition to micro polarization curves and the extrapolation in the Tafel plot to overpotentials equal to zero was done and the results are in good agreement. The values of j_0 in literature show a large scatter between less than $1mA/cm^2$ and several $100mA/cm^2$. This was obtained with different techniques such as rotating disc electrode and scanning electrochemical microscopy ranging from single crystal surfaces to Pt nanostructures in different surfaces [4, 5, 18, 103, 134-139]. Extracting a reliable value for j_0 from all these different references is still a challenge.

2.5.2 Gold

As compared to platinum and palladium, gold is also not affected by air and water and hardly assailable by acids or alkalis. Au thin films on mica support are often used in electrochemistry instead of single crystals with comparable results [51, 140]. A thin Au film is evaporated on the mica with a very thin chromium interlayer to achieve a good adhesion. Flame annealing is the common procedure to prepare thin Au films and Au single crystals that lead to large (111) oriented terraces.

Cyclic voltammograms for Au(111), Au(100) and Au(110) surfaces in 0.1 M H_2SO_4 are shown in [105]. Especially the investigations in sulfuric acid with the typical current peaks in the voltammograms are an indicator for the single crystal quality. Surface reconstruction of Au(111) and Au(100) is lifted at potentials positive of 0.58V versus NHE due to adsorption of sulfate ions [141]. This leads to the unreconstructed (1x1) surface. Gold oxide formation is beginning at potentials higher than 1.24V versus NHE. Due to the importance of Au(111) in electrochemistry as support material some important properties will be summarized. Au (111) has several advantages such as easy preparation process and large potential window in the double layer region. The quality of Au(111) single crystals can be easily checked by cyclic voltammetry in sulfuric acid and evaluating the peaks attached to the lifting of reconstruction and forming ordered ($\sqrt{3}\times\sqrt{7}$) R19.1° sulfate layer [105, 142-144]. Defects as well as small terraces will inhibit the well pronounced peaks at 0.58V and 1.02V versus NHE for the above mentioned procedures. Details on investigations on other low index Au single crystal surfaces can be found in [141, 145-151]

The investigations on Au single crystals towards HER are quite rare [152-154] and show a weak dependence on crystallographic orientation in early studies. Comparable to Pt precise investigations can only be done with high quality of single crystals and high standards for clean experimental conditions. Perez et al. [99] investigated the hydrogen evolution in different Au(hkl) single crystal electrodes. To avoid a strong influence of adsorbing ions and to correlate the structure sensitive reactivity to the different single crystal electrodes, perchloric acid was used as electrolyte. Using a rotating disc setup with hanging meniscus, concentration gradients of hydrogen at the interface were avoided. Similar results to the above mentioned results on Pt crystal surfaces were obtained. Comparable to Pt the different Au(hkl) surfaces are also structure sensitive. The hydrogen evolution is distinctly different on the three Au(111), Au(100) and Au(110) surfaces regarding the HER [99]. The electrocatalytic activity is dependent on the crystallographic orientation of Au single crystal surfaces in the following sequence Au(111) > Au(100) > Au(110). It was suggested that HER activity increases with the atomic density of the surface following the trend of increasing work function for the electron [155]. While the activity for hydrogen evolution is also dependent on the crystallographic orientation seen in the case of Pt the measured net current densities are several orders of magnitude lower for Au.

2.5.3 Palladium

Palladium, similar to gold and platinum, is easily deformable and drawable to nearly any kind of profile such as gold and platinum. It is not affected by air and water but dissolves in oxidizing acids and molten alkalis. A unique property of Pd is the high hydrogen absorption ability which may even change the lattice constant. This behavior complicates the preparation procedure because any kind of hydrogen during annealing would negatively influence the quality. Due to the hydrogen absorption into the bulk Pd the hydrogen adsorption current region is superposed with the hydrogen absorption current in the cyclic voltammogram. In order to avoid this problem, thin Pd films which were epitaxially grown on supports such as Au(hkl) and Pt(hkl) were used to study the hydrogen adsorption as well as the hydrogen evolution and several other reactions [16, 44, 46-48, 156-163]. As shown in [105] for Pd(111) the potential window of the double layer region is large and typical current peaks for hydrogen adsorption and oxide formation are pronounced. As mentioned before Pd is often used as thin film on different supports, hence the most experimental findings are dealing with Pd overlayers. Unexpected results for the hydrogen reactions of thin Pd layers were obtained towards the electrocatalytic activity which will be reported in later sections.

2.5.4 Rhodium

Rhodium is used as catalyst material which is unaffected by acids but attacked by molten alkalis. Compared to Pt and Au flame annealing is a typical procedure to obtain well-ordered Rh surfaces when cooling in a H_2/Ar-mixture [164, 165]. Also cooling down in CO leads to well-ordered Rh surfaces whereas an oxidation of the CO layer without structural changes is not possible. Cyclic voltammograms for well-ordered Rh(111) in can be found in [105, 166, 167]. Typical peaks in the CV in the hydrogen adsorption region at NHE are overlapped with anion adsorption in sulfuric acid. Whereas in the broad double layer region, the sulfate ion coverage is constant [168]. Wan and Itaya [169] also observed a (3x 7)R19.1° structure of sulfate using STM. Due to the strong interaction of Rh surfaces with anions the reduction of perchlorate is was described in detail by Rhee et al. [170] and investigated in detail on Rh(111) in perchloric acid by Clavilier et al. [164] Up to now there is almost no electrochemical investigation of Rh(100) and Rh(110) whereas Rh(111) is often used in STM and compared with Pt(111) [169, 171].

2.5.5 Iridium

Ir has similar physical and chemical properties when compared to Pt. In electrochemistry Ir single crystals were used for the first time by Motoo and Furuya [172]. The electrodes were annealed in a propane-oxygen flame at ca. 2000 °C for 5 minutes and cooled down in pure hydrogen. The results showed that the currents in the hydrogen adsorption region was strongly dependent on the crystallographic orientation and on the supporting anions in the electrolyte [173]. The low index faces Ir(110) and Ir(100) are reconstructed in UHV. The (100) facet forms a surface with a (5x1) superstructure [174, 175]. As in the case of Au(110) and Pt(110) a missing row type in form of a (1x2) superstructure is observed in the Ir(110) reconstructed surfaces [176]. The formation of an ordered ($\sqrt{3}$x $\sqrt{7}$)R19.1° sulfate adlayer [169] was also observed as in the case of Pt(111) [123, 129]. Typical CVs of Ir(111) [177, 178] and Ir (100) [164, 172, 177, 179] are reported in literature.

2.5.6 Ruthenium

Ruthenium as a member of so the called platinum group has a hexagonal closed packed (hcp) crystal structure which is in contrast to all other investigated surfaces with face centered cubic (fcc) crystals structures. Cyclic voltammograms of a well prepared Ru(0001) electrode in perchloric acid are shown in [105]. Ru is used as co-catalyst material for platinum anodes in the low temperature range. Single crystalline Ru(0001) surfaces are therefore used in several studies as model surface to a obtain a basic understanding of catalytic processes [180-185] whereas the preparation was here done in UHV. Clean and well ordered surfaces were also prepared by El Aziz and Kibler [186] using an inductive heating in an argon stream. Detailed electrochemical studies by CO displacement ascribed a peak in the hydrogen evolution region to OH reduction and hydrogen adsorption [186]. This peak is dependent on the pH and on the presence of specifically adsorbing anions on the surface. Cyclic voltammograms of Ru single crystal surfaces are also given in literature [105, 187-189] with different electrochemical focuses.

2.5.7 Highly Oriented Pyrolytic Graphite (HOPG)

Highly oriented pyrolytic graphite (HOPG) is sp-2 hybridized carbon and characterized by the highest degree of three-dimensional ordering. The density, parameters of the crystal lattice, the (0001) orientation and anisotropy are physical properties which are close to those for natural graphite. HOPG consists of stacked planes composed to a lamellar structure. Interaction of carbon atoms within a single plane is based on covalent bonds and therefore stronger compared to those from adjacent planes. A single-atom thick plane is called graphene. The lattice of graphene consists of two equivalent interpenetrating triangular carbon sublattices A and B. Within a plane each carbon atom is surrounded by three nearest neighbors forming a honeycomb structure. Atomic resolved HOPG with a STM normally shows a close packed array where each atom is surrounded by six nearest neighbors. The distance between two atoms is 0.246 nm. However, under ideal conditions the true structure of graphene the hexagonal rings with an atomic distance of 0.1415 nm can be resolved. HOPG can serve as an ideal atomically flat surface to be used as a substrate or calibration grid for SPM investigations. The higher the quality, the less the roughness of the surface and the smaller the number of steps and defect sites, steps can be separated by several µm. As an electrochemically inert support with high overpotentials for hydrogen and oxygen evolution it is often used as carbon support for model system studies [20, 190-192].

2.6 Deposition Techniques

Electrochemical as well as local methods for nanostructuring enable modifications of the substrate on a defined position by direct impact. Underpotential and Overpotential deposition as well as metal nucleation and growth are standard methods to decorate surfaces. The local approach of formation of nanostructures which is caused by macroscopic techniques is called top-down technology. With this technique it is possible to create defects or to deposit clusters on the surface. The structures are directly fabricated in contrast to using a mask for indirect deposition.

2.6.1 Underpotential and Overpotential Deposition of Metals

For the deposition of a foreign metal Me on a metal substrate S a potential shift towards positive potentials can occur compared to the expected deposition potential from the Nernst equation. This phenomena is called underpotential deposition (UPD) and is usually limited to one ML [193]. The bulk metal deposition appears below the equilibrium potential and is called overpotential deposition (OPD). In typical cyclic voltammograms the UPD is shown by additional adsorption and desorption peaks. These peaks strongly depend on the quality of the substrate and the crystallographic orientation. Different UPD systems at single crystals are given in the literature [193, 194]. Polycrystalline substrates often show not so well pronounced UPD peaks, which is probably caused by different crystallographic areas and more surface imperfections (e.g. steps, kinks, vacancies) in comparison to single crystals.

2.6.2 Metal Nucleation and Growth

The deposition of a foreign metal Me$_{ads}$ on a metal substrate S is determined by two factors: a) the binding energy between Me$_{ads}$-Me$_{ads}$ and Me$_{ads}$-S and b) the crystallographic misfit between the foreign metal Me$_{ads}$ and the substrate S. This misfit is defined by Budevski et al.[193] as:

$$mf = \frac{d_{0,Me} - d_{0,S}}{d_{0,S}}$$

$d_{0,Me}$ and $d_{0,S}$ are the interatomic distances of the deposited metal Me$_{ads}$ and the substrate S. Depending on these parameters, there are three idealized growth mechanisms (see Figure 2.7):

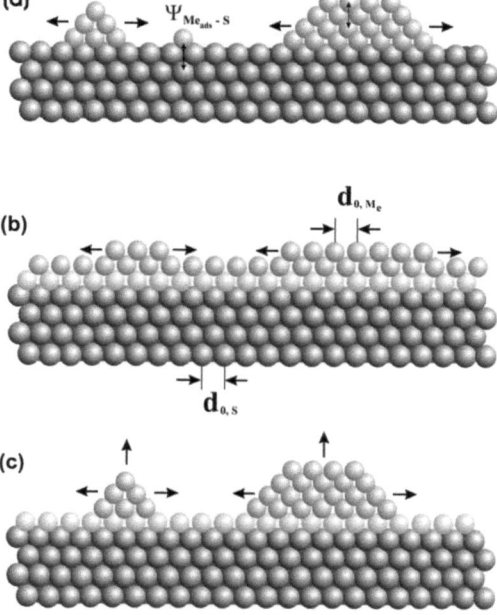

Figure 2. 7 Different growth mechanisms from [193]; (a) Volmer-Weber mechanism with 3D island growth; (b) Frank-van der Merwe mechanism with one monolayer in the UPD regime and layer by layer growth in the OPD regime; (c) Stranski-Krastanov mechanism with one monolayer in the UPD regime and 3D island growth on top of the first deposited monolayer.

a) If the binding energy $\psi_{Me_{ads}-Me}$ between Me$_{ads}$ and Me is higher than the binding energy $\psi_{Me_{ads}-S}$ between Me$_{ads}$ and S there is no effect by the crystallographic misfit. This growth mode is called Volmer-Weber mechanism. This mode is distinguished by the creation of 3D-bulk islands.

b) If the binding energy $\psi_{Me_{ads}-Me}$ between Me$_{ads}$ and Me is lower than the binding energy $\psi_{Me_{ads}-S}$ between Me$_{ads}$ and S and the misfit is negligible, the deposition process is called Frank-van der Merwe mechanism. Because of the strong interaction between Me$_{ads}$ – S interaction there is a layer by layer growth in the first and is followed by a continues growth like on a native Me substrate.

c) If the binding energy $\psi_{Me_{ads}-Me}$ between Me$_{ads}$ and Me is lower than the binding energy $\psi_{Me_{ads}-S}$ between Me$_{ads}$ and S and there is significant misfit, the deposition process is called Stranski-Krastanov mechanism. In this model the first layer of the adsorbed metal has a different strain compared to the bulk Me material. After the first 2D layer growth there is a 3-D crystallites growth on top of the deposited layer.

2.6.3 Single Particle Deposition

Different techniques for nanolithography such as the STM and the EC-STM are discussed in the following. In electrochemical environment there are a lot of different options to use the STM tip to create a cluster or induce a defect. The first attempt uses the tip and forces a contact with the substrate, see Figure 2.8 (a). Thereby defect for metal deposition are created. A more gentle approach is shown in Figure 2.8 (b) which is quite sensitive to the substrate. The tip removes only a tarnishing film from the substrate and allows local metal deposition. Studies at the system Cu on Au with sodium dodecylsulphate films are shown in [195]. The double layer cross talk in Figure 2.8 (c) has to be included if the electrochemical double layer of the tip and the substrate interferes. A loss of the potential control of the substrate underneath the tip can be caused. Different groups investigated the Cu/Cu^{2+} [196] and the Ag/Ag^{+} [197] system where dissolution of bulk metal was induced by a very close approach of the tip. Schuster [198] and Kirchner et al. [199] used a different way to create structures, the so called electrochemical machining. They applied an ultra short pulse between tip and substrate in nanosecond regime for etching processes. Due to the time dependence of the double layer charging on the upper or lower part of the tip one can use this technique to set the pulse in such a way that only the tip apex interacts with the substrate potential Figure 2.8 (d). So it is possible to dissolve metal by oxidation in a small area. An easy way creating tip induced nanostructures is the burst like method in Figure 2.8 (e). The metal on the tip is deposited directly under the tip [200, 201]. Schindler and co-workers [202] explain the effect with a high metal ion concentration after the deposition process which shifts the Nernst potential for the surface under the tip. Figure 2.8 (f) shows the scanning electrochemical microscopy (SECM). Next to the previously mentioned techniques the SECM can be used for the detection of electrochemical processes on a micrometer scale. Especially in the field of electrocatalysis the SECM is a powerful tool for direct investigations of particles. More details will be found in [203, 204]. Most previous works reported in literature deal with the deposition of clusters on substrates induced by the jump-to-contact method Figure 2.8 (g). The tip gets in contact with the substrate without destroying it. This performs a formation of a metal bridge, so called connective neck, which breaks if the tip is retracted. Different groups use this technique to create single particles as well as large arrays from different metals on various substrates [76, 205-208].

In this work massive metal tips will be used without loading the tip with another metal from a solution. The tips will consist of platinum, palladium and gold for deposition of single nanoparticles, whereas the jump to contact method is used.

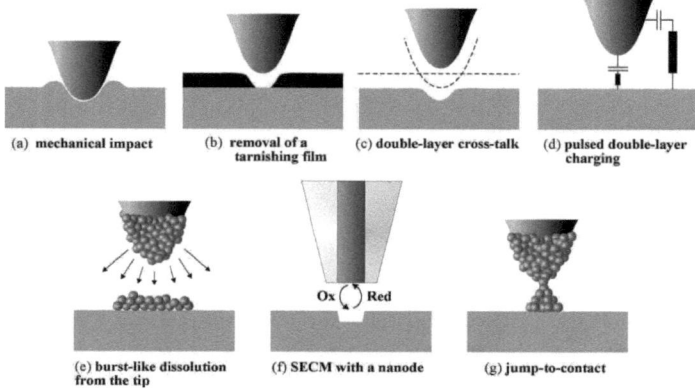

Figure 2. 8 Different methods for nanostructering with an STM tip from [70].

2.7 Hydrogen Evolution / Hydrogen Oxidation Reaction (HER)/(HOR)

The hydrogen evolution reaction (HER) is intensively investigated by experimental as well as theoretical research groups due its simplex reaction pathway consisting of three fundamental reactions:

- Discharge reaction of a proton to form an adsorbed hydrogen atom, known as Volmer reaction [209]:

$$Pt + H^+ + e^- \rightarrow H - Pt \quad (2.17)$$

- Combination of two adsorbed hydrogen atoms to form molecular hydrogen, known as Tafel reaction [210]:

$$2(H - Pt) \rightarrow H_2 + 2Pt \quad (2.18)$$

- Combination of an adsorbed hydrogen atom with a proton and an electron to form molecular hydrogen, known as Heyrovsky reaction [211]:

$$H - Pt + H^+ + e^- \rightarrow H_2 + Pt \quad (2.19)$$

resulting in the overall reaction:

$$2H^+ + 2e^- \rightarrow H_2 \qquad (2.20)$$

This reaction can occur in two different reaction pathways, the Tafel-Volmer [209, 210] and the Heyrovsky-Volmer [209, 211] pathway. Molecular hydrogen is formed by the recombination of two adsorbed hydrogen atoms in the Tafel-Volmer reaction. The Heyrovsky-Volmer reaction describes the formation of molecular hydrogen from one adsorbed hydrogen atom and one proton from the electrolyte which is discharged at the adsorbed hydrogen atom on the surface. Therefore the adsorption of atomic hydrogen is the important issue. These elementary steps are also consequential for the hydrogen oxidation reaction (HOR).

The electrocatalytic activity can be described by the exchange current density j_0, which is dependent on the electrode material and the concentration of educts and products. A classical representation is the plot of $\log(j_0)$ *versus* the adsorption energy ΔG_{ad} of the hydrogen to the metal resulting in the so called Volcano plot [212]. Both branches of the plot end in a vertex where $\Delta G_{ad} \approx 0$ [213]. Results from experimental and theoretical groups substantiate the relationship of $\log(j_0)$ and ΔG_{ad} in the Volcano plot [8, 14, 16] (Figure 2.9). Typically, noble metal catalysts such as Pd, Pt, etc. or alloys of these metals are located near the top of the volcano. They exhibit a neither too strong nor too weak interaction of the respective metal with hydrogen ($\Delta G_{ad} \approx 0$). One theoretical study was done in the group of Nørskov [8, 10] using density functional theory (DFT) calculations demonstrated a volcano type behavior of hydrogen chemisorption energies in respect to the exchange current density for hydrogen evolution.

Hydrogen reactions were also experimentally investigated on polycrystalline Pt as well as single crystal surfaces in a wide range. Markovic et al. [96, 97] and Barber et al. [98, 100] found that the reactivity of different Pt(hkl) surfaces towards hydrogen reactions is strongly influenced by surface orientation. Another approach is decorating supports with foreign metals, which leads to surprising findings and deeper understanding of catalytic understanding. This occurs if catalytic active material such as Pt and Pd are deposited onto foreign metal supports. The reactivity towards the hydrogen reactions is different compared to the bulk material due to lattice strain of the Pt or Pd overlayer. Adsorption and desorption processes of hydrogen were investigated on different decorated single crystal surfaces in the group of Kolb [13, 44, 46]. Various indicators were found which underline the importance of overlayer thickness and selection of support. Overlayer thickness as well as the supporting metal influence the adsorption and desorption behavior and thus electrocatalytic activity. The experimental finding was supported by theoretical work from the group of Nørskov [30, 33, 34], the group of Groß [11, 12, 37] and the group of Schmickler [38-42]. Experimental results from Hernandez and Baltruschat [6, 7] indicate the importance of step sites for the electrocatalytic reactivity of Pd-layers on vicinally stepped gold electrode surfaces. The hydrogen adsorption and HER were studied on Au single crystals with different step densities modified with Pd. Markovic and Ross [133] studied the system Pd on Pt(111). Here, also an improvement of the catalytic activity towards the HER and HOR compared to pure Pt(111) was found. A variation of the thickness was not done, but the positive effect of Pd as overlayer influenced by the support was also demonstrated in this case. Meier et al. [19, 21] investigated the local reactivity at single Pd particles on Au(111) surfaces. It was shown that with decreasing height of single Pd nanoparticles on Au(111) the electrochemical activity towards hydrogen evolution reaction (HER) is increased by more than two orders of magnitude. This result is comparable to the increased catalytic activity for HER on Pd nanostructured Au(111) in the work of Kibler [17] and Pandelov and Stimming [16]. Also Hernandez and Baltruschat [6, 7] reported enhanced activity of Pd step decorated Au(332) surfaces. An enhanced mass transport on ultra microelectrodes [18, 214] and highly dispersed particles in carbon supports [4, 5] caused also higher activity compared to standard systems. The experimental results from all groups mentioned above were ascribed to several effects regarding electronic and geometric factors such as a strain effect in the lattice [33, 35], high activity of low coordinated atoms [6, 7, 215], a direct involvement of the gold surface [15] and enhanced mass transport [4, 5, 18, 214]. However, a complete understanding that describes the overall effect is not found until now. For a good comparison of all mentioned results a direct access to reliable values of j_0 for hydrogen related reactions seems to be a great challenge [17]. Although there are a number of experimental techniques and theoretical approaches, values of j_0 differ considerably in literature [4, 5, 18, 101]. Nevertheless, due to the high catalytic activity Pt surfaces are of special interest as catalysts whereas Au as an inert material regarding hydrogen catalysis compared to Pt is a suitable support. However, a complete understanding that describes the overall effect is not found until now.

A suitable value to compare the activity of the different systems is the exchange current density j_0 which describes the electrocatalytic activity of the surface and is dependent on the surface structure as well as the concentration of educts and products. Conway and Bockris [216] reported a correlation of the exchange current density j_0 to the electronic work function φ which defines the binding energy between electrons near the Fermi level and the material. For HER it was also shown that the logarithm of j_0 increases as the d character of the material increases. The latter was explained by the fact that as the d character increases, more electrons have paired spins and hence they require more energy to extract them resulting in a decrease of ΔH of adsorbed hydrogen atoms. Parsons [213] studied the relationship between exchange current density and the ability of the electrode to adsorb atomic hydrogen in terms of the standard free energy ΔG_H. His theoretical studies showed that $\log j_0$ reaches a maximum when $\Delta G_H \approx 0$. Metals such as Pt, Pd, Rh, and Ir that adsorb moderately hydrogen have high values of j_0.

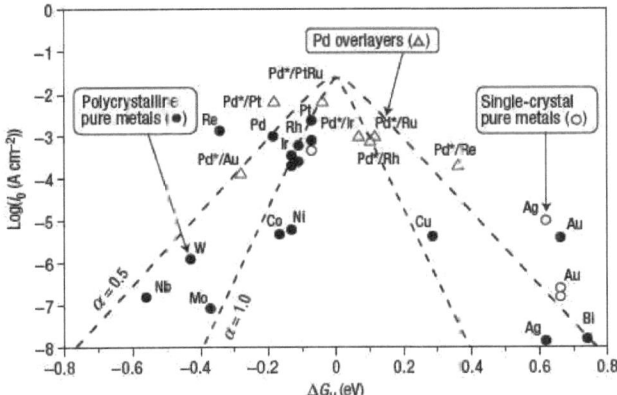

Figure 2. 9 Volcano plot for the HER for various pure metals and metal overlayers from [8].

Trasatti [212] obtained two fairly parallel lines for the exchange current density j_0 being plotted versus the work function. As mentioned previously the rate and mechanism of HER and HOR depends on the bond strength between metal and hydrogen atom (M-H). A pass through a maximum is suggested and a similar volcano shape curve was reported by Krishtalik and Delahay [217]. Pt is close to the top of the volcano curve where the Pt-H bond is neither too strong nor too weak. A similar study was done by Nørskov et al. [8, 10], see Figure 2.10. Density functional theory (DFT) calculations demonstrated a volcano type behavior of hydrogen chemisorption energies with respect to the exchange current density for hydrogen evolution. Pt was found again to be a better catalyst than other metals for HER primarily because hydrogen evolution reaction on Pt is thermo-neutral at the equilibrium potential. The findings of this work can be used to predict the behavior of other bimetallic systems for HER as well as HOR. The analysis was reported as a new method to obtain H adsorption free energies and understand trends for different systems that are of electrochemical interest.

2.8 Oxygen Reduction Reaction (ORR)

The oxygen reduction reaction (ORR) is one of the most central reactions due to its importance in e.g. fuel cell applications and metal air battery systems. A direct reaction pathway occurs via a four electron process:

$$O_2 + 4H^+ + 4e^- \rightarrow 2H_2O \qquad U_{00} = 1.23 \text{ V} \quad \textbf{(2. 21)}$$

An indirect pathway occurs via hydrogen peroxide in two two-electron processes:

$$O_2 + 2H^+ + 2e^- \rightarrow H_2O_2 \qquad U_{00} = 0.682 \text{ V} \quad \textbf{(2. 22)}$$

$$H_2O_2 + 2H^+ + 2e^- \rightarrow 2H_2O \qquad U_{00} = 1.77V \quad \textbf{(2. 23)}$$

In the latter case hydrogen peroxide is formed as intermediate, which partly diffuses in the electrolyte phase. During the last decades several different models were developed to describe the interplay between the two parallel reaction pathways (detailed review [218]). A simplified reaction scheme for the oxygen reduction reaction is shown in [219], based on the reaction schemes by Wroblowa et al. [220] and Bagotskii et al. [221]. The scheme distinguishes between the preadsorbed oxygen being reduced through a direct four electron reduction resulting in the formation of water or through a series pathway forming H_2O_2 as an intermediate which is further reduced to water. In addition hydrogen peroxide can either chemically decompose to water and O_2 or desorbs into the solution. The different reaction pathways are interpreted as the consequence of different adsorption states. The assumed adsorption possibilities for oxygen are shown in detail in [218] whereas possible reaction pathways are shown in [218, 222]. One-sided adsorbed oxygen prefers the pathway via H_2O_2. Adsorption on both sides stretches the O-O-binding and thus decomposes of the O_2 molecule. Next to these models the reaction pathway also depends on the pH of the electrolyte, the presence or absence of adsorbing anions and on the catalyst material. A series reaction pathway is observed on graphite, most carbon materials as well as gold and mercury. Parallel existence of both the direct and the series pathway is found on platinum, platinum alloys, palladium, Pt-like metals, and silver. Heterogeneous decomposition of peroxide is obtained with platinum, silver, spinels, and perovskites. Generally in alkaline electrolyte the series pathway is enhanced compared to the direct pathway. It is worth mentioning that in an acidic surrounding, carbon materials do not show satisfactory catalytic activity for either of the pathways [222, 223].

The oxygen reduction was well investigated on single crystal surfaces of various metals [133, 224, 225]. Investigations of the electrocatalytic activity of nanostructured and bimetallic surfaces are also reported in an extended manner [47, 48, 226-232], but a comprehensive understanding is still missing. Therefore, ORR electrocatalysts with defined parameters are necessary to comprehend the origin of the catalytic properties. It is well described in [133, 224-226] that crystallographic orientation, morphology, and chemical composition are correlated to the catalytic activity. Grgur et al. [226] reported an increased oxygen reduction activity from (111) < (100) < (110) orientation of Pt single crystals in aqueous sulfuric acid. Compared to pure platinum crystals a new possible approach for effective catalysts is to implement 3d metals into the bulk or on the surface. Bimetallic systems such as Pt-Fe alloys were reported for the first time from Toda et al. [230], here the ORR activity is influenced by a so called Pt skin effect. The activity for ORR was found under certain circumstances to be twenty-five times higher compared to pure Pt. It was shown that the surface of such alloys consists of pure Pt (Pt skin layer) and the reactivity decreases with increasing thickness of the skin layer. This effect is explained by a modified electronic structure of the skin layer compared to bulk Pt.

Stamenkovic et al. [233-235] suggested bimetallic systems such as Pt_3Ni, Pt_3Ti, Pt_3Fe, Pt_3Co etc.. Due to a change in the electronic properties of Pt, the metal oxygen binding energy to bimetallic surfaces can be advantageously modified. These results were explained by different states of adsorbed oxygen resulting in different activation energies. Paulus et al. [227] reported experimental results which show an increase of the catalytic activity per Pt atom for different Pt alloys. Investigations by Schmidt et al. [236] showed an enhanced ORR electrocatalytic activity with Pd modified Au and Pt single crystals Naohara et al. [47, 48] found a similar behavior for the ORR for thin Pd overlayers on Au single crystals. Pd monolayer and alloy catalysts were also investigated by Shao et al. [228, 237]. He found higher catalytic activity for alloys and Pd monolayer supported on Ru, Rh, Pt and Au. The origin of this effect is described by a d-band centre shift of the deposited overlayer [237, 238]. Greeley et al. [239] combined in a recent study theoretical predictions of the experimental results by investigating Pt alloyed with transition metals such as Sc and Y. A positive effect on the ORR activity, showed by positive half wave potential shift of up to 60mV, was found for polycrystalline Pt_3Sc and Pt_3Y electrodes which were also calculated to be a stable binary alloy.

Besides the field of platinum-based alloys, several platinum free materials are known to show (partly considerable) activity towards oxygen reduction. These catalysts were mainly investigated in the context of methanol crossover studies guided by the desire to find an ORR catalyst which exhibits a high tolerance towards simultaneously present methanol or other organic impurities on the cathode side. One example is rhodium, investigated as part of a transition-metal sulfide [240], as rhodium-rhodium oxide catalyst [241] or rhodium sulfide catalyst [242]. The latter two catalysts were studied for their application as oxygen reduction catalysts in industrial HCl electrolyzers where the hydrogen-evolving cathode is substituted by an oxygen-consuming cathode. Other examples are metal-containing porphyrin and annulene systems [243-248]. It is, however, not only challenging to overcome the high activation losses of the ORR. Additionally the ORR is highly sensitive to the presence of every kind of impurity, even of traces of impurities (organic, anionic and cationic impurities or additives). A full discussion of the poisoning is hardly possible. An overview/review is given "Poisons for the O2 Reduction Reaction" and the references cited therein [249].

Recent work on batteries, especially metal air systems also pointed out the importance of the oxygen side. Recent publications underline the importance of noble metal electrodes made of gold or platinum or alloyed particles from gold and platinum [250, 251].

2.9 Methanol Oxidation Reaction (MOR)

The electrooxidation of methanol on platinum surfaces in acidic media has been experimentally and theoretically studied in great detail in the last decades [133, 252-266] and can be written in the overall reaction as:

$$CH_3OH + H_2O \rightarrow CO_2 + 6H^+ + 6e^- \qquad (2.24)$$

The adsorbed methanol is stepwise dehydrogenated resulting in the formation of adsorbed CO_{ads}. Besides adsorbed CO also the existence of the intermediate COH_{ads} is assumed.

$$CH_3OH \rightarrow CH_3OH_{ads} \qquad (2.25)$$
$$CH_3OH_{ads} \rightarrow CO_{ads} + 4H^+ + 4e^- \qquad (2.26)$$

The adsorbed CO is oxidized via OH_{ads}, which is formed by the dissociative adsorption of water.

$$H_2O \rightarrow OH_{ad} + H^+ + e^- \qquad (2.27)$$
$$CO_{ad} + OH_{ad} \rightarrow CO_2 + H^+ + e^- \qquad (2.28)$$

In parallel to these main reaction pathways side reactions like the build up of formic acid [258, 267] are also possible. But until now the elementary steps are still under investigation although a lot of studies for methanol oxidation on well defined Pt and Pt alloy surfaces discussing the overall pathways and reaction mechanisms were performed [133, 257, 260, 268-274]. The oxidation of methanol is following a dual reaction pathway via an indirect or a direct reaction to form CO_2. The indirect pathway is determined by the formation of adsorbed CO which is then oxidized to CO_2. Depending on the potential the adsorbed CO can block the water activation at low applied potentials. Higher potentials lead to an oxidation of CO to CO_2. The direct pathway proceeds via the formation of formaldehyde and formic acid which are soluble in the electrolyte and can be oxidized to form CO_2. Both intermediates can subsequently be oxidized to CO_2. As already mentioned above, the electrocatalysis of methanol has challenging tasks to solve. CO is a strong adsorbate and a stable intermediate during the methanol decomposition. Especially the strong poisoning of the common used platinum containing electrodes in fuel cells leads to an decreasing activity [275]. The anode overpotential in a direct methanol fuel cell (DMFC) with state of the art technology is higher compared to the overpotential in a hydrogen fuel cell (PEM-FC). This results in reduced overall cell efficiency although the thermodynamic standard potentials of the hydrogen oxidation and the methanol oxidation are similar. However, the methanol oxidation is by several orders of magnitude slower compared to the hydrogen oxidation.

Solving these tasks implies modifying the catalyst on which the methanol oxidation occurs. Although platinum is the catalyst with the highest reactivity towards MOR the overpotential for adequate activity is until now very high for applications with high power output. For the well known Pt-Ru alloy [276, 277] and its derivates [278-280] a lower overpotential and an increased reactivity was found, explained by ligand [281] or bifunctional [282, 283] effects. However, there are a lot of improvements to do. The activity is not as high as demanded and the compositions which are under current investigation are consisting of expensive and rare metals.

3 Materials and Methods

3.1 Chemicals

Chemicals	Formula	Provider	Purity
Perchloric Acid	$HClO_4$	Merck	suprapur
Sulfuric Acid	H_2SO_4	Merck	suprapur
Sulfuric Acid	H_2SO_4	Merck	p.a.
Hydrochloric Acid	HCl	Merck	p.a.
Hydrogenperoxide	H_2O_2	Merck	p.a.
Sodium Perchlorate Monohyd.	$NaClO_4$	Merck	p.a.
Potassiumtetrachloroplatinate	K_2PtCl_4	Aldrich	99.995%
Potassiumhexachloroplatinate	K_2PtCl_6	AlfaAesar	99.9%
Palladium(II) nitrate	$Pd(NO_3)_2$	MaTeck	p.a.
Palladium sulfate	$PdSO_4$	AlfaAesar	99.95%
Copper(II) sulfate hydrate	$CuSO_4$	AlfaAesar	99.999%
Copper(II) perchlorate	$CuClO_4$	AlfaAesar	99.999%
Acetone	C_3H_6O	Sigma Aldrich	p.a.
Isopropanol	C_3H_8OH	Sigma Aldrich	p.a.
Sodium thiocyanate	$NaSCN$	BDH Prolabo	p.a.
Potassium hydroxide	KOH	BDH Prolabo	p.a.
Argon	Ar	Linde	4.8
Hydrogen	H_2	Linde	5.0
Oxygen	O_2	Linde	5.0
Carbonmonoxid	CO	Linde	3.7
Milli-Q water	H_2O	18MΩcm, 3ppm TOC	

Nobel metal wires such as gold (Au) with a diameter of 0.25 mm and a diameter of 0.5 mm, palladium (Pd, diameter 0.25 mm), platinum (Pt, diameter 0.25 mm) and platinum/iridium (Pt/Ir with a ratio Pt:Ir of 80:20, diameter 0.25 mm) were purchased from Carl Schäfer GmbH Co.KG (Germany). The Au, Pd and Pt wires had a purity of 99.9%. The metal wires were cleaned in Caro's acid and were flame annealed in a Bunsen burner before each experiment.

3.2 Electrochemical Setup

3.2.1 Electrochemical Glass Cell

All electrochemical measurements were preformed in glass cells in a standard three electrode arrangement consisting of the working electrode (WE), at which the electrochemical process of interest occurs, the reference electrode (RE) and the counter electrode (CE). The glass cell consists of two compartments where the reference electrode is separated from the working electrode and the counter electrode. In order to minimize the potential drop in the electrolyte, the reference electrode is placed very close to the working electrode through a Luggin capillary. The upper part of the cell and the electrolyte can be purged with inert gas through two gas inlets made of glass tubes in order to remove the oxygen from the electrolyte. The outgoing gases passes through a small glass reservoir filled with water (the bubble counter), in order to prevent the diffusion of air back into the cell. Commercial mercury/mercury sulfate electrodes (Hg|Hg$_2$SO$_4$, in 0.1M H$_2$SO$_4$, B 3610, Schott, Germany) were used as reference electrodes (RE) in all experiments with a constant potential of 660 mV vs. NHE. For all experiments the counter electrode (CE) consisted of a gold mash or sheet which was connected to a gold wire and melted in a glass grinding.

3.2.2 Potentiostats

Cyclic voltammetry experiments and galvanostatic pulse measurements were performed with a software controlled potentiostat/galvanostat either HEKA PG 310 (HEKA Elektronik Dr. Schulze GmbH) with the data acquisition software PotPulse v. 8.77 or an Autolab PGSTAT 30 (Metrohm Autolab B. V., Netherlands) with the data acquisition software GPES. A potentiostat maintains the applied potential to the working electrode in an electrochemical cell with respect to the reference electrode. Changes in the electrochemical conditions cause a current flow between the working electrode and the counter electrode in order to keep the potential constant. Vice versa a galvanostat keeps the current between working electrode and counter electrode constant by adjusting the necessary potential of the working electrode.

3.3 Electrochemical and SPM Techniques

3.3.1 Setup EC-SPMs, (STM, AFM, SECPM)

In this thesis two different electrochemical scanning probe microscopes (SPM) were used. One EC-STM was a homebuilt setup consisting of a Nanoscope IIIA controller (Veeco Instruments Inc., USA) and a PicoSPM STM base (Agilent Technologies former Molecular Imaging) together with an EC-Tec bipotentiostat/galvanostat BP600 and an EC-Tec bi-scangenerator SG600. The STM data were recorded with the computer software Nanoscope v 4.43r6 whereas electrochemical data were recorded with a labview program BP600 or a digital phosphor oscilloscope Tektronix TDS5034B (Tektronix, USA). In this setup the tip is attached to the scanner (so called tip-scanner) and scans the fixed sample [284] The sample is mounted on a ferromagnetic sample holder which is magnetically connected via three micrometer screws with the STM base below the scanner. The EC cell consists of a Teflon ring with a diameter of 5 mm and an electrolyte volume of 100µl and is pressed on the sample. Two screws serve for a first manual approach of the STM tip to the sample surface. The third screw is computer controlled by a stepper motor and is therefore responsible for the fine adjustment and the final approach of the tip. After each step the measured tunneling current is compared to the setpoint value. As long as the measured current is lower than the chosen setpoint value the z-piezo continues extending until the tunneling conditions selected in the software are established. The scanner has a maximum scan range of 5µm x 5µm in xy-direction and 2µm in z-direction. The used current/voltage converter has a sensitivity of 10 nA/V.

The other SPM was a commercial combined in situ EC-AFM/EC-STM/SECPM instrument consisting of an electrochemical Veeco Multimode system with the Veeco universal bipotentiostat, optional a combined STM/SECPM head or an AFM head, a Nanoscope III D Controller and Nanoscope 5.31r2 or 6.12 software which provides SPM and EC data acquisition. In this setup the tip is fixed in position. It is placed above the sample which sits on top of the scanner, i.e. the sample scans in xyz-direction, so called sample-scanner. The working electrode is positioned between a ferromagnetic plate and a Teflon cell (diameter 5 mm, electrolyte volume 100µl). Throughout the experiments a scanner type E with a range of (10µm x 10µm x 2.5µm) in xyz-direction was utilized. The SECPM head amplifies the measured potential difference between the potential controlled substrate electrode and a tip which is at open-circuit with four gain values (1, 10, 50, 100) and a leakage current of only a few fA. In STM mode the tunneling current is converted by 10nA/V.

3.3.2 Tip Preparation

STM tips from various metals were produced to create local nanostructures. Tips from gold, palladium and platinum wires were etched with different etching solutions and different etching parameters. All wires have a diameter of 0.25mm. The setup used for the electrochemical etching process is shown in Figure 3.1. The wires are passed through a platinum ring (10mm in diameter and 0.5mm thick Pt wire) and adjusted in the centre of the ring to achieve a symmetrical apex form. Gold and palladium tips were etched with +1.65V dc voltage between the wire and the etching electrode in a hydrogen chloride lamella.

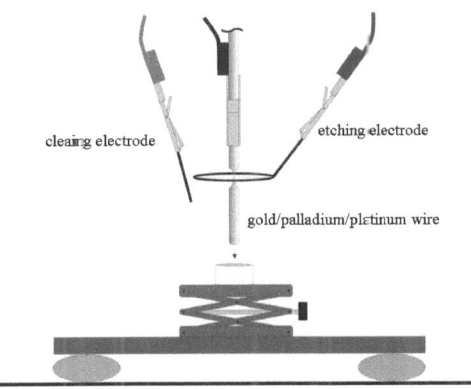

Figure 3. 1: Electrochemical etching set up from [285].

Freshly etched tips were immediately stored in hot Milli-Q water to remove the rests of oxides from the tip. After the process the etching wire was cleaned by switching the voltage to -5V between etching electrode and cleaning electrode and put in hydrogen chloride solution to remove the oxides. Typical times for the etching process were 2min for the gold wire and 5min for the palladium wire to achieve good results. For the etching procedure of platinum an ac voltage with a dc offset has to be used. Simply dc voltages would not be able to do electrochemical etching with platinum. Therefore, $6.5V_{pp}$ ac voltage with a 1.6V dc offset is applied between the platinum wire and the etching ring and uses 5M KOH + 3.5M NaSCN electrolyte. Electrochemical polishing of the Pt tips was used to smooth the surfaces and sharpen the tips. The best parameters found here were cycling between -4V and +4V vs. NHE in $1M\ H_2SO_4 + 1M\ HNO_3 + 1M\ H_3PO_4$ with a sweep rate of 1V/s for ten times.

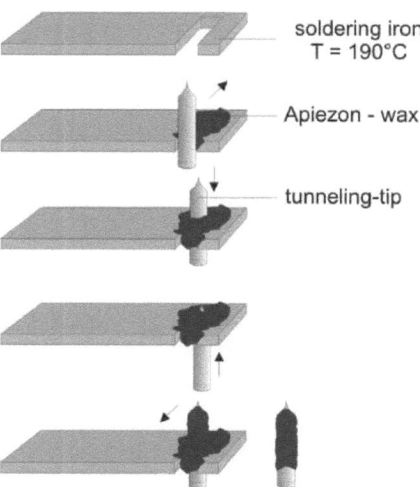

Figure 3. 2: Schematic draw of EC-STM tip insulation with hot Apiezon wax from [285]; the arrows indicate the direction of the movement of the tip.

EC-STM tips require a chemical resistant insulation to minimize the Faradaic currents at the tip. These currents, caused by electrochemical reactions, would overlap the tunneling current if the tip would not be insulated. A lot of different insulation techniques such as electrophoretical paint, Apiezon wax, nail polish, glass coating, etc., were used. In this work the tips were insulated with Apiezon wax. Using a soldering iron at about 200°C the wax was melted and the tip was passed through the hot wax by using micrometer screws. After the whole tip was dipped into the wax a small free area at the apex broke open (see Figure 3.2). In Figure 3.3 different STM/SECPM tips are shown. Gold as well as Palladium tips have a high reproducibility and easy etching procedure. Gold tips with diameters smaller than 40nm and Palladium tips with diameter of smaller than 100nm were prepared. An easy etching procedure with simple dc voltage and hydrogenchloride solution allows a reproducible method to create high quality nanoelectrodes. First tests with an optical microscope to check the apex form if it is sharp and conical allow a fast determination for further use or elimination. Long and thin tips as well as mechanical damaged tips were eliminated after the optical check. Random samplings were imaged in the SEM to obtain a higher resolution and to control the etching parameters. If too many tips show a bad performance, the set up was upgraded (cleaning, change of etching wire, diameter of etching wire, etc.).

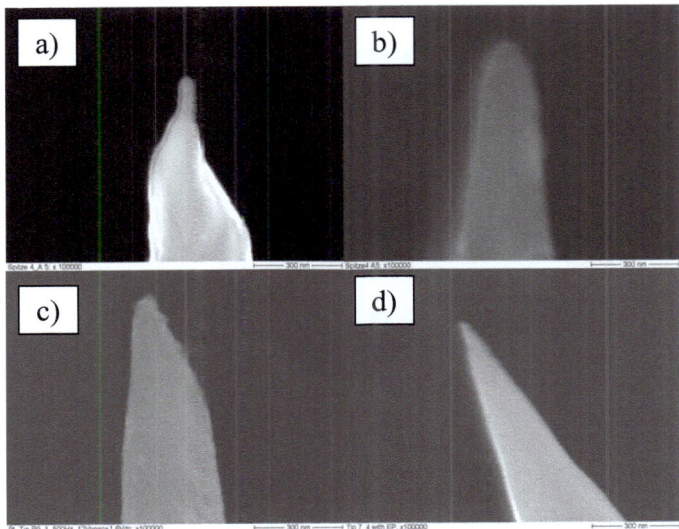

Figure 3. 3: scale bar = 300nm; a) Au tip, b) Pd tip, c) Pt tip after etching, d) Pt tip after electrochemical polishing.

3.3.3 Reference Electrodes for EC-SPM

Due to the limited space available in the SPM setups wire electrodes are required. For all EC SPM measurements gold/gold oxide or platinum/platinum oxide wire electrodes were used as reference electrodes. Before and after each experiment the potential of the electrodes was measured versus a mercury/mercury sulfate electrode stored in 0.1 M H_2SO_4. The gold/gold oxide and the platinum/platinum oxide reference electrodes showed a stable equilibrium potential behavior However, one has to keep in mind that the potential behavior of these reference electrodes is pH sensitive. Therefore, the equilibrium potential of the electrodes was measured versus a commercial mercury/mercury sulfate electrode in different concentrations of $HClO_4$ using a salt bridge which minimizes diffusion potentials. A slope of 58 mV per change of one pH and an equilibrium potential U_{00} = 1.32 V vs. NHE for the gold/gold oxide electrode was measured. The platinum/platinum oxide electrode had a slope of 57mV per change of one pH and an equilibrium potential of U_{00} = 0.93 V vs. NHE. Both results are in good agreement with the Nernst equation and the literature data of the equilibrium potential [286].

3.3.4 Local Reactivity Measurements

The basic idea of local reactivity measurements is to use the EC-STM tip as a local sensor to measure currents. The reaction products of an active species on the working electrode are reversibly converted at the tip into a Faradaic current. Here the hydrogen evolution reaction (HER):

$$2H^+ + 2e^- \rightarrow H_2$$

at the particle is investigated whereas the reverse reaction hydrogen oxidation (HOR)

$$H_2 \rightarrow 2H^+ + 2e^-$$

is detected with the STM tip, schematically shown in Figure 3.4.

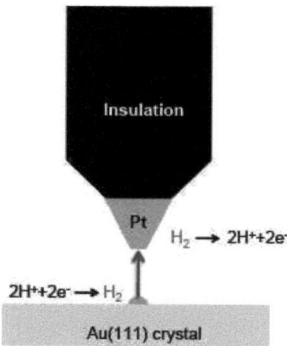

Figure 3. 4: Schematics of the used method. Evolved hydrogen at the catalyst particle is oxidized and thus detected with the STM tip.

A sequence of steps has to be done in order to perform local reactivity measurements:

1st: The tip is directly placed over the bare gold surface or a created particle to perform the reactivity measurement. Therefore, the STM feedback is turned off and the tip is retracted 25 nm. The tip current now only consists of the Faradaic current, no tunneling current or overlapping double layers can interfere.

2nd: The potential of the working electrode is varied to evolve hydrogen at the particle, which is then recorded as hydrogen oxidation current at the tip. A one second long negative potential pulse is applied to the working electrode, the tip potential is kept constant at 350mV vs. NHE and the tip current is recorded.

3rd: After the potential pulse, the tip stays retracted for another 35 seconds to record the complete hydrogen oxidation current.

4th: Again the tip is approached towards the surface and another morphology image of the cluster is recorded. Then the tip is set once more above the cluster, the procedure starts again at 1 with a different overpotential for hydrogen evolution reaction.

During the whole measurement procedure the tip potential is kept constant at 350mV vs. NHE. A typical graph of the resulting current transients is shown in Figure 3.5, all recorded with an oscilloscope. The black curve indicates the measurements on a bare gold surface I_{Au}, whereas the upper curve was recorded on top of a cluster $I_{Pt\ on\ Au}$ that was created. The blue curve shows the difference of the red and black curve, which indicates the hydrogen oxidation current arising from the particle I_{Pt}.

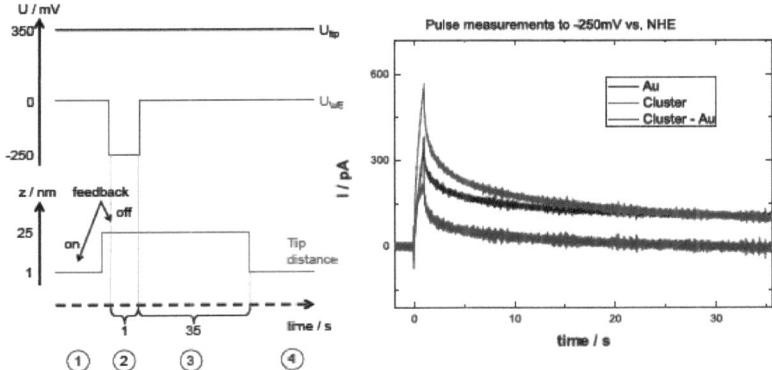

Figure 3. 5: Left: Scheme of the potential distribution at the tip (black) and the working electrode (grey) together with the tip distance / nm (light grey); right: Typical current transient of local reactivity measurements, showing reactivity of bare gold (dark grey), after creation of a cluster (light grey) and the corrected current evolved by the particle (black).

The hydrogen oxidation current is in first order proportional to the concentration of hydrogen in solution and therefore a rate for the activity of the platinum particle on the gold surface. The evaluation procedure of the recorded current transients is as follows: The current from the single Pt particle I_{Pt} (sum current minus background, $I_{Pt\ on\ Au} - I_{Au}$) and the area that is integrated to gain the charge that got transposed by hydrogen oxidation at the tip. This charge, typically values of several 100pAs, then is divided by the pulse time, here one second, to get an average current through the particle. The current density j is derived dividing the integrated corrected current values by the surface of the cluster, which were in the range between 20 and 200 nm² obtained from STM images. With this value and the Butler-Volmer equation with a transfer coefficient of 0.5 [16] for HER and HOR and standard values for the gas constant R, T and F, the exchange current density j_0 of a single particle can be calculated.

3.3.5 Potential Behavior of Pd/H Electrodes – pH Dependency

The primary standard for pH measurements is the reversible hydrogen electrode (RHE), Pt/H$_2$, H$^+_{aq}$ [287, 288]. According to the Nernst equation (equation 5.1) its potential is given by

$$U_0 = U_{00} + \frac{RT}{2F}\ln f_{H_2} - \frac{RT}{F}pH \qquad (5.1)$$

where the pH represents the proton concentration, U_{00} is the standard potential, by convention 0 mV at all temperatures and f_{H_2} is the fugacity of hydrogen. However, for analytical applications of the

RHE the electrolyte must contain hydrogen of defined partial pressure. Whereas, in contrast to platinum palladium, also being a member of the platinum group metals, possesses the ability to store hydrogen within its crystal lattice and is therefore able to function as a pH sensitive palladium hydride (Pd/H) [95, 289, 290] and can be also used in hydrogen free solutions [291]. The pH sensitivity of palladium hydride can be understood by the palladium/hydrogen phase diagram [290]. At room temperature three different phases can exist depending on the atomic ratio of H/Pd [288, 292]:

1) H/Pd ≤ 0.025: low hydrogen containing phase where the activity (or chemical potential) of hydrogen varies depending on the amount of hydrogen in the α phase. When the hydrogen content is increased to the edge of the miscibility gap, the system is said to have only one degree of freedom (at T = const), which fixes the hydrogen content and therefore, the hydrogen activity of the phase.

2) 0.025 ≤ H/Pd ≤ 0.6: intermediate miscibility gap consisting of two phases, α and β phase. According to equilibrium the activity of hydrogen in the α phase must be equal to the activity in the β phase, i.e. within this region of hydrogen content the activity of hydrogen remains constant providing a wide-ranging and stable hydrogen activity against which the hydrogen ion activity in solution can be reliably measured. The potential of 50 mV vs. RHE was attributed to a mixed potential due to the presence of the two-phase region with the α phase predominating [293].

3) H/Pd > 0.6: high hydrogen-containing β phase, the potential drops to zero volt.

The potential of a Pd/H electrode is dependent on composition in the pure α and β phases, but is independent of composition in the mixed α + β phase region. This means using a Pd/H tip electrode in the miscibility gap, a change in the proton concentration changes the equilibrium potential of the Pd/H electrode.

3.4 Electrodes and Electrode Preparation

3.4.1 Au(111) Single Crystal and Au(111) Films on Mica

Au(111) electrodes were prepared from two different kinds of material, sputtered Au(111) substrates (Metallhandel Schroer GmbH) and Au(111) single crystals (MaTecK GmbH). The substrates consist of borosilicate glass support sputtered with 2.5nm chromium layer and 250nm gold layer, whereas the chromium is needed to have a better adhesion of the gold. After cleaning for at most one minute in Caro's acid the sputtered substrate was rinsed with Milli-Q water and afterwards flame annealed with a Bunsen burner. Typically the gold (substrates as well as single crystals) was heated until a red glow was visible. This temperature was kept for some minutes and a cooling down process followed. The procedure was repeated several times and results in large atomically flat terraces and the typical gold 60° triangle structure. The crystal was cleaned by an oxidation process in 1M H_2SO_4 or 1M $HClO_4$. Afterwards the crystal was stored 5 minutes in hydrogen chloride to remove the gold oxides and contaminations. Flame annealing with the Bunsen burner and cooling down in H_2 atmosphere for several times reorganized the Au(111) structure and results in a homogeneous and atomically flat surface.

3.4.2 HOPG

Prior to each experiment the surface of the HOPG electrode (ZYB, Mikromash) was cleaved with an adhesive tape. All experiments were performed at the basal plane of the HOPG surface. In order to have a well-defined surface area and to avoid the involvement of the edge plane in electrochemical experiments, the electrode was partly covered with Teflon tape (CMC Klebetechnik, Germany) exposing a geometric surface area of 0.2 cm^2 to the electrolyte.

3.4.3 Ru(0001), Rh(111), Ir(111) and Ir(100) Single Crystalline Films

All single crystalline surfaces were used as received from the UHV preparation chamber from the group of M. Schreck, Uni Augsburg. Impurities due to transport and storage were cleaned in an ultra sonic bath. Therefore, the electrodes were put in acetone, isopropanol and Milli-Q water, according to this sequence, each for 5 minutes and afterwards they were again carefully rinsed with Milli-Q water. Storage and transportation afterwards was done with a droplet of ultrapure water to protect the surface. For electrochemical experiments the electrode was partly covered with Teflon tape (CMC Klebetechnik, Germany) exposing a geometric surface area of 0.2 cm^2 to the electrolyte.

3.5 Deposition Techniques

3.5.1 Single Particles with an STM

As described in Chapter 2, different methods for local nanostructures are possible. In this thesis the jump to contact technique using STM was used to create nanoparticles on Au(111). For different deposited metals (palladium, platinum, gold) different pure metal tips consisting of Pd, Pt, Au and Pt loaded Au tips were used. This former technique allows a direct use of the etched metal wires in the EC-STM system and saves a predeposition with foreign metals on the tip by electrochemical methods as shown in Chapter 2. After insulation of the different metals with Apiezon wax, they were used in the EC-STM setup.

Different methods allow achieving a contact between the tip and the substrate to form a connective neck and deposit tip metal on the substrate [205, 208]. Meier et al. [284] applied a potential pulse to the z-piezo to achieve a displacement. Jung [294] used a similar technique and added Visual Basic scripts to generate particle arrays. In this work a software controlled technique was used to operate the movement of the tip. This nanoscript lithography software controls of the x, y and z direction of the tip, time delays, geometrical forms, analogue outputs controls etc. Hence different programs were developed to generate single particles and particle arrays. The only request for a single particle is to stop the imaging process of the EC-STM, switch off the feedback and move the tip towards the substrate. Particle arrays were generated in a similar way. The imaging was stopped, the feedback was switched off and the program for nanolithography was started.

The potential of the working electrode was always kept below the equilibrium potential of the deposited metal to avoid desorption.

3.5.2 Electrochemical Metal Deposition from Solution

The pulse deposition procedure starts at an initial potential of 860mV $vs.$ NHE. The potential is then switched to the final deposition potential of 660mV $vs.$ NHE for different periods of time, in order to vary the amount of noble metal deposited. 0.1M $HClO_4$ solution containing 0.5mM $Pd(NO_3)_2$ for all Pd deposition experiments was used. Further measurements were also performed in 0.1M/1M $HClO_4$. Platinum deposition took place in the same way but with a 1M $HClO_4$ containing 0.5mM K_2PtCl_4 or 0.5mM K_2PtCl_6. Further measurements were also performed in 1M $HClO_4$. After the noble metal deposition the substrate was rinsed several times with Milli-Q water to avoid spontaneous Pd or Pt deposition on the substrate. By integrating the current transient during the deposition the Faradaic charge of the deposited metal can be calculated taking into account the double layer charging. The two electron reduction process from Pd^{2+} to Pd results in 420 μCcm^{-2} for one monolayer of Pd which is the same for Pt^{2+}. For the couple Pt^{4+} to Pt the reduction processes via four electrons and results in a charge of 840 μCcm^{-2} for a full monolayer of Pt. For an additional determination of the amount of noble metal hydrogen adsorption was used. The hydrogen adsorption method deals with a typical charge value of about $210\mu Ccm^{-2}$ for an upd layer of adsorbed hydrogen. The evaluation was done by integration of current transients in the hydrogen adsorption region. The amount of deposited Pt was additionally investigated with CO stripping experiments, whereas a value of $280\mu Ccm^{-2}$ for the oxidation of one monolayer of CO was used. STM images, evaluated with WSxM 4.0 (Nanotec Electronica S.L.) [295], showed the morphology of the particles size and distribution and allow a determination of the deposited metal. The deposition of Pd was always done using a single potentiostatic pulse method. Pt was deposited via single and double potentiostatic pulses. As mentioned above a variation of time leads to a variation of the amount of deposited noble metal.

4 Results

4.1 EC-SPM on Au(111), HOPG, Ru(0001), Rh(111), Ir(111), Ir(100)

In order to study substrate effects in electrocatalysis different supports were investigated. As well known surfaces HOPG and Au(111) were used to calibrate the SPMs and to supply reference electrodes which were already accurately investigated for electrochemical purposes. HOPG and Au(111) were also used as a benchmark for the new SECPM technique to compare to well established methods such as STM and AFM. Ru, Rh and Ir as new single crystalline surfaces were investigated to check their quality and their suitability for electrochemical investigations and especially for providing new supports which may lead to new experimental findings.

4.1.1 HOPG and Au(111)

HOPG (0001) has been the support of choice for many STM experiments as it provides atomically flat areas, good mechanical stability and electronic conductivity. Figure 4.1 A1 and Figure 4.1 A2 show a combined EC-STM and SECPM study of a freshly cleaved HOPG surface. Due to the possibility to switch between EC-STM and SECPM during scanning one and the same (500 nm x 500 nm) surface area was first scanned in constant current mode EC-STM (Figure 4.1 A1) and afterwards in constant potential mode SECPM (Figure 4.1 A2). The EC-STM image shows the characteristic atomically flat and large terraces with a typical well pronounced step edge across the section. On well prepared samples such steps are observed every few micrometers. According to literature the nominal spacing between single graphene layers is 3.35 Å [296] therefore, a monoatomic HOPG step has a height of 3.35 Å and multiples of this value for multilayered steps. The height of the step edge observed in Figure 4.14 A1 is evaluated to be 1.049 nm which is equal to a spacing of three graphene layers. The atomic topography of HOPG usually shows a hexagonal close-packed (hcp) lattice with a lattice constant of 0.246 nm [297]. The inset in Figure 4.1 A1 represents atomically resolved HOPG with a lattice constant of 0.248 nm which is in line with literature values. However, the measured nearest neighbor distance is 0.124 nm indicating that not only the top layer, but also the second layer of the graphite crystal is resolved. Imaging the second graphene layer requires a sharp STM tip geometry and is usually not achieved in SPM experiments. Due to its high conductivity all HOPG surface features can be sharply resolved by STM.

While the STM technique is based on the tunneling effect, SECPM, which is also a tool to image surfaces under electrochemical conditions uses the potential difference between tip and electrode as feedback signal to control the tip sample distance. As it can be seen in Figure 4.1 A2 the SECPM is able to image all features with a similar resolution compared to the STM. SECPM maps a smooth graphite surface along the terraces indicating a homogenous potential distribution. However, the step edge shows a much higher image contrast and appears broader than monitored by STM. Furthermore, damped oscillations can be observed in direct vicinity of the step edge. This observation suggests a significant change in the electrochemical properties, probably in the electrochemical surface potential at the step edge. Therefore, this phenomenon is under further investigation due to the possibility to investigate irregularities at the HOPG steps possibly resulting from changed binding conditions compared to surface atoms. Neglecting the oscillations SECPM reveals a step height of 1.084 nm which is still in good agreement with a step over three monolayers.

In the past, intensive studies have shown that defect sites on basal plane HOPG play an important role especially in influencing and controlling the electrochemical behavior of the basal plane [298, 299]. The anisotropy of HOPG leads to several problems when examining parameters such as adsorption, capacitance and electron transfer kinetics which often vary with the proportions of edge and basal plane on the exposed surface. Based on $Fe(CN)_6^{-3/-4}$ studies, Rice and McCreery [300] found a 1000 times larger electron transfer rate for the basal plane when only 0.1 % of edge defects, with a transfer rate of 0.1 cms^{-1}, are present on the basal plane, with a transfer rate of 10^{-7} cms^{-1}. Furthermore, defect sites such as steps and edges have dangling bonds, giving rise for charged functional groups and reactive sites. Such disordered structures have a locally increased density of states due to the variety of energy levels created by defects, functional groups and local dipoles [301]. Whereas macroscopic techniques such as differential capacitance measurements only yield a spatially averaged response over the whole surface, SECPM is able to map the local potential distribution. Therefore, SECPM provides to image structural heterogeneities of the basal plane with high spatial resolution and is a promising tool for further investigations of the electrochemical properties and the surface chemistry of HOPG defect sites. Figure 4.1 A3 shows a contact AFM image of 2 µm x 2 µm obtained in air with atomically resolved HOPG in the inset (0.243 nm). The flat HOPG surface is imaged with a resolution comparable to STM and SECPM under electrochemical conditions. The step density is very low which emphasizes the homogeneity of the (0001) graphite surface over a large area showing only one step edge per micrometer.

Due to the fact that the atomic resolution can easily be obtained by STM (Figure 4.1 A1 Inset) and AFM (Figure 4.1 A3 Inset) HOPG is used as a calibration grid for the piezos in x-y direction in SPM. However, atomic resolution has not yet been achieved by SECPM.

In order to test the resolution of the SECPM on metal surfaces a Au(111) single crystal was examined. Figure 4.1 shows a typical in situ EC-STM image of an Au(111) single crystal in 0.1 M

HClO$_4$ taken at a potential of 0.5 V vs. NHE. The image was obtained in constant current mode applying a bias voltage of +100 mV and a tunneling current of 1 nA. The scanned area of the gold surface shows atomically flat (111) terraces separated by well-defined monoatomically high gold steps. A line scan analysis of the EC-STM image reveals a step height of approximately 2.4 Å which is in good agreement with the theoretical value of 2.35 Å. Furthermore, the angle of the Au(111) terrace edge in the middle of the picture is 58.6 °; 120/60 ° step edges are characteristic of the Au(111) orientation. These results were directly compared with scanning electrochemical potential microscopy (SECPM). The same area was scanned in constant potential mode applying a potential difference of 5 mV between the tip and the Au(111) single crystal electrode. As can be seen in Figure 4.1 B2 no significant differences between the STM and the SECPM image are observed. The angle of the gold terrace in SECPM is 62.9 ° and the line scan analysis reveals an average step height between 2.4 and 2.5 Å showing the high quality imaging with this SPM technique.

Figure 4.14 B3 shows a 2 µm x 2 µm overview image of Au(111) sputtered on glass obtained in contact mode AFM showing grain boundaries of the (111) fibre textured gold film. The image shows the deflection of the cantilever signal. These defect sites were not obtained using a massive Au(111) single crystal in the case of STM and SECPM investigations.

It was possible to resolve the atomic structure of the electrode surface applying STM (Figure 4.1 B1 Inset) and AFM (Figure 4.1 B3 Inset). The observed nearest neighbor atomic spacing is about 0.286 nm in STM and 0.276 nm in AFM. Taking the inaccuracy of the techniques into account both values are in good agreement with the literature value of the fcc lattice constant of 0.289 nm of the Au(111) surface [302].

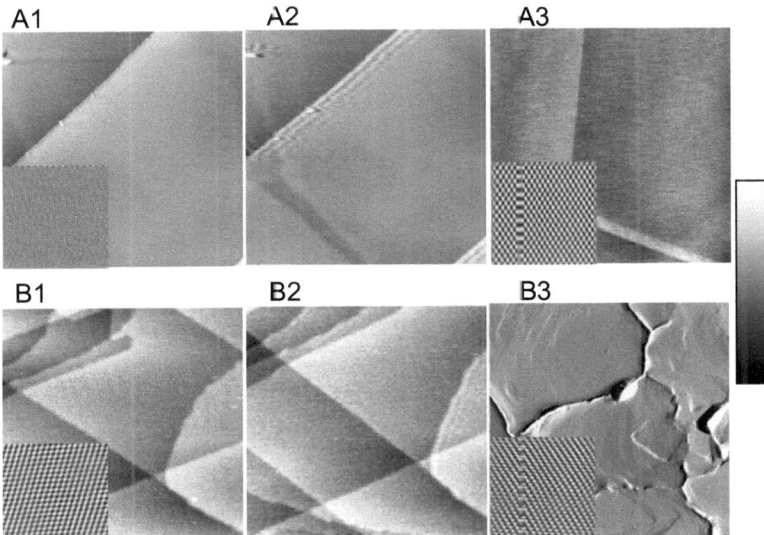

Figure 4. 1: EC-STM, SECPM and AFM of HOPG (A) and Au(111) (B) single crystalline electrode: (A1) EC-STM (500 nm x 500 nm, h_{max} = 3.68 nm, Inset: 5 nm x 5 nm), U_S = 700mV vs. NHE (A2) SECPM (500 nm x 500 nm, h_{max} = 3.89 nm) image of HOPG in 0.1 M $HClO_4$ at U_S = 700 mV vs. NHE, (A3) Contact mode AFM in air (2 µm x 2 µm, h_{max} = 3.3nm), (B1) EC-STM (150 nm x 150 nm, h_{max} = 1.5 nm, Inset: 5 nm x 5 nm) U_S = 500mV vs. NHE, (B2) SECPM (150 nm x 150 nm, h_{max} = 1.5 nm) image of Au(111) single crystal electrode in 0.1 M $HClO_4$, U_S = 500 mV vs. NHE; (B3) Contact Mode AFM in air of a (111) fibre textured gold film (2 µm x 2 µm, U_{max} = 0.1V, Inset: 5 nm x 5 nm). Imaging conditions: STM: I_T = 1 nA, U_{Bias} = +100 mV, SECPM: ΔU = 5 mV. The insets show atomic resolution.

4.1.2 Ru(0001)

Figure 4.2 shows the new Ru(0001) single crystalline surface in EC-STM (Figure 4.2 A), SECPM (Figure 4.2 B) and AFM (Figure 4.2 C) down to atomic resolution (Inset Figure 4.2 C). The Ru(0001) structure is characterized by large atomically flat terraces with an average width of up to 100 nm. The terraces were resolved in a similar quality by all three SPM techniques independent of the environmental conditions, i.e. whether recorded in air or in electrolyte. The EC-STM and SECPM images were obtained in 0.1 M $HClO_4$ applying an electrode potential of 500 mV vs. NHE. While the EC-STM image shown in Figure 4.2 monitors sharp edges of the single crystalline Ru(0001) structure these edges appear fringed in the SECPM image (Figure 4.2 B). These results indicate that the electronic properties of the step edges, i.e. conductivity mapped by STM and the potential distribution mapped by SECPM may be different from the electronic properties of the smooth terraces. Similar phenomena were already observed for the HOPG(0001) surface where the defect site rich steps lead to an inhomogeneous potential distribution and therefore, the steps have a much higher image contrast in SECPM than in EC-STM (see Figure 4.1 A1 and Figure 4.1 A2).

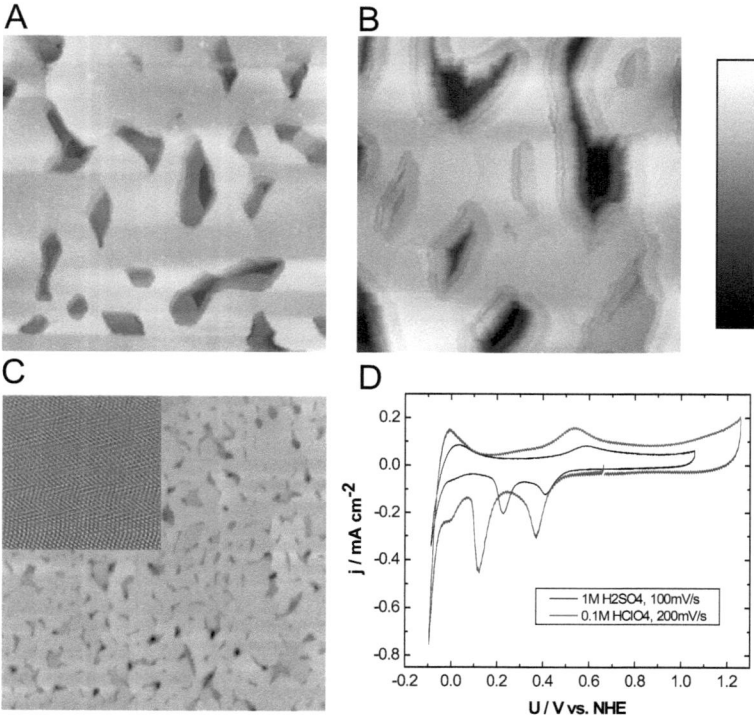

Figure 4.2: EC-STM, SECPM, AFM and CV of Ru(0001): (A) EC-STM (500 nm x 500 nm, h_{max} = 12.17 nm), U_S = 500mV vs. NHE, (B) SECPM (500 nm x 500 nm, h_{max} = 17.22 nm) image of Ru(0001) in 0.1 M HClO$_4$ at U_S = 500 mV vs. NHE, (C) Contact mode AFM in air (5 μm x 5 μm, h_{max} = 40 nm, Inset: atomic resolution, 12 nm x 12 nm) and (D) CVs obtained in 1 M H$_2$SO$_4$ (thin black curve) and 0.1 M HClO$_4$ (black curve) with a scan rate of 100 mVs^{-1} and 200 mVs^{-1}, respectively. Imaging conditions: STM: I_T = 1 nA, U_{bias} = 100 mV, SECPM: ΔU = 5 mV.

Ruthenium crystallizes in a hexagonal close-packed lattice with a lattice constant of 0.271 nm [303, 304]. Applying contact mode AFM it was possible to resolve the atomic structure shown in the inset in Figure 4.2 C with a measured next neighbor distance of 0.27 nm. This value is in good agreement with the literature data reported above. However, AFM was performed in air at room temperature. Under these conditions oxygen is chemisorbed on the ruthenium surface. LEED experiments and DFT calculations have shown that the O adlayers on Ru(0001) can vary between low coverage up to a full monolayer (Θ = 1) occurring at high gas partial pressure [305, 306]. Depending on the coverage the structure of the O adlayer forms a (2 x 2) or (2 x 1) phase, for high surface coverage, i.e. Θ = 1 the O adatoms sit at the hcp-hollow sites with a (1 x 1) periodicity adopting the lattice constant of the underlying (0001) facet. Assuming that the Ru(0001) single crystalline support is completely covered by oxygen under ambient conditions it is not clear whether the atomic resolution seen in Figure 4.2 C shows oxygen or ruthenium atoms. However, the observed hcp structure represents the (0001) facet of the ruthenium support.

Electrochemical characterization on Ru(0001) was done via cyclic voltammetry in sulfuric and perchloric acid of different concentrations. The CVs of Ru(0001) obtained in 1 M H_2SO_4 (black curve, Figure 4.2 D) and 0.1 M $HClO_4$ (black curve, Figure 4.2 D) were recorded with a scan rate of 100 mVs^{-1} and 200 mVs^{-1}, respectively, show well pronounced peaks. The formation of ruthenium hydroxide starts at different potentials depending on the electrolyte and is clearly seen in the anodic peaks at around 0.6 V vs. NHE. The formation of RuO_2 at potentials higher than around 1 V vs. NHE was avoided [307]. The two sharp cathodic peaks at 400mV vs. NHE are at similar potential the more negative ones differ by 100 mV. Both cathodic peaks were ascribed to represent the OH reduction and the hydrogen adsorption. A specific adsorption of anions is also discussed although perchlorate ions are known to be weakly adsorbing ions compared to sulfate ions. Sulfate ions are accountable to protect Ru(0001) surfaces from oxidation at lower potentials while supporting a faster reduction process of ruthenium hydroxide [189]. A strong dependence of the perchlorate concentration on the hydrogen adsorption was found in our investigations according to [186]. All observed features are also reported on Ru single crystals surfaces in literature [105, 187-189] showing the high quality of the single crystalline surfaces for electrochemical investigations.

4.1.3 Rh(111)

Figure 4.3 shows the EC-SPM and the cyclic voltammetry study of Rh(111). The EC-STM (Figure 4.3 A) and SECPM (Figure 4.3 B) images were obtained in 0.1 M $HClO_4$ applying an electrode potential of 500 mV vs. NHE. Both images show a closed rhodium layer with a regular pattern of the typical (111) triangular surface structure. The (111) terraces have an average width of 80 nm. Step heights are evaluated by a detailed line scan analysis; 0.210 nm for EC-STM (Figure 4.3 A) and 0.222 nm for SECPM (Figure 4.3 B). These values are in good agreement with the Rh(111) monoatomic step height which can be calculated by the lattice constant a of bulk rhodium according to $h = 3^{-1/2}$ a. Taking the theoretical (a = 0.383 nm) [308] and experimental (a = 0.380 nm) [309] fcc lattice constant of rhodium into account the step height is 0.221 nm and 0.219 nm, respectively. While the EC-STM image reveals the typical 120/60 ° step edges of the (111) structure with an accuracy of ± 1 °, the SECPM monitors the Rh(111) structure slightly distorted. The angles are different than 60 °, e.g. α has a value of 62.8 °, whereas β accounts only 53.8 °. These results probably arise from scan artifacts due to the necessary slow scan rate in SECPM mode due to a stronger influence of thermal drift compared to the higher scan rate in STM.

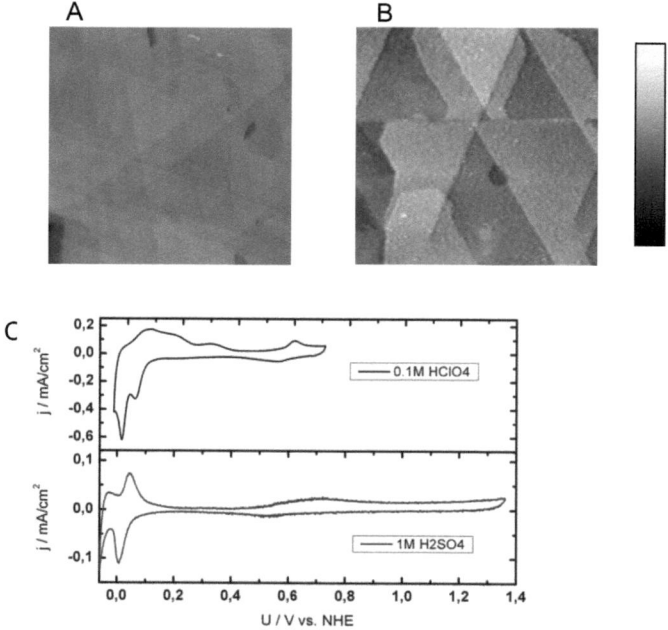

Figure 4. 3: EC-STM, SECPM and CV of Rh(111): **(A)** In situ EC-STM (500 nm x 500 nm, h_{max} = 10 nm) U_S = 500mV vs. NHE, **(B)** SECPM (200 nm x 200 nm, h_{max} = 4 nm) and image of Rh(111) in 0.1 M HClO$_4$ at U_S = 500 mV vs. NHE **(C)** CVs obtained in 0.1 M HClO$_4$ (black curve) and 1 M H$_2$SO$_4$ (thin black curve) with a scan rate of 100 mVs^{-1}. Imaging conditions: STM: I_T = 1 nA, U_{bias} = 100 mV, SECPM: ΔU = 5 mV.

Rh(111) single crystalline surfaces were investigated in sulfuric and perchloric acid solutions of different concentrations. For direct comparison the CVs in 1 M H$_2$SO$_4$ (thin black curve) and 0.1 M HClO$_4$ (black curve) are shown in Figure 4.3 C. Due to the difficulties in Rh single crystal preparation [310] there are only few reports in literature compared to work on for example Pt and Au single crystals. The CV in sulfuric acid (red curve in Figure 4.3 C) has three characteristic regions. The hydrogen adsorption and desorption is well pronounced indicated by the sharp pair of peaks which are observed before the hydrogen evolution. At more positive potentials the surface is covered with a ($\sqrt{3}$ x $\sqrt{7}$) (hydrogen-) sulfate adlayer [165, 166] which was resolved by in situ electrochemical STM by Wan et al.[169] is similar to the work of Funtikov et al. on Pt(111) [123, 129]. At potentials more positive than 0.6 V vs. NHE a formation of surface oxides on terraces can be observed represented by broad peaks. All specific peaks are comparable to the work of Sung et al. [166] and Xu et al. [311]. A CV of Rh(111) in perchloric acid is also shown a in Figure 4.3 C (black curve) which is slightly different from that obtained in sulfuric acid (red curve) and peak sharpness is not as clear as on massive single crystals [164, 169]. The most important difference is the second peak in the hydrogen region. According to Clavelier et al.[164] the reduction of perchlorate ions and the adsorption of reduction products take place at potentials slightly positive from the hydrogen adsorption/desorption or even overlaps.

4.1.4 Ir(111) and Ir(100)

Ir behaves similar to Pt resulting in related physical and chemical properties. Here we show SPM and electrochemical studies of single crystalline Ir surfaces with (111) and (100) orientation. As it can be seen in Figure 4.4 A1 and A2 typical (111) oriented surfaces were imaged with SECPM and AFM. In a scan area of 500 nm x 500 nm Ir(111) shows well defined triangle structures with extended surfaces. The typical angle of 60 ° was only observed at an imaging scan rate larger than 1 Hz in scan areas smaller than 50 nm x 50 nm. Due to the necessary scan rate smaller than 0.2 Hz in the SECPM mode in images such as Figure 4.4 A1 the 60° was not achieved. This can be ascribed to scan artifacts due to thermal drifts causing piezo movements which are more noticable at slower scan rates. Contact mode AFM images with a higher scan rate are resulting in a higher precision regarding thermal effects as can be seen in Figure 4.4 B1. The holes indicated by the dark brown color in both images show defect sites from the growth process. Evaluating the average step height leads to a value between 0.24 and 0.25 nm for SECPM measurements. STM images reveal a step height of 0.23 nm. The observed value is in line with literature of 0.221 nm which was calculated from the fcc lattice constant of 0.383 nm [309]). These results indicate the high precision of the different SPM systems. Also SECPM is able to resolve this in vertical direction as accurately as the STM.

Ir(100) was also investigated with electrochemical and SPM methods. A typical image, obtained by STM in air, is shown in Figure 4.4 B1. It is clearly seen that the structure is completely different compared to the (111) oriented one. For an overview a 500 nm x 500 nm picture is shown. The defect density is not as large as for Ir(111) resulting in a smooth surface with a RMS value of 1 nm. Well pronounced 90° angles are visible with terrace widths of up to 50 nm.

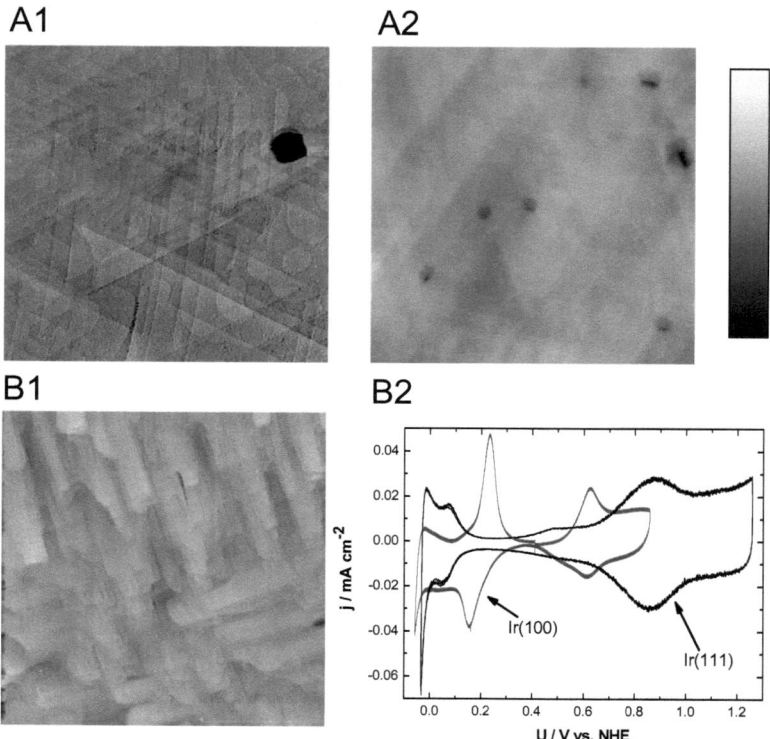

Figure 4. 4: SECPM, AFM and CV of Ir(111) and Ir(100): (A1) SECPM image of Ir(111) (500 nm x 500 nm, h_{max} = 5.5 nm) in 0.1M $HClO_4$ at U_S = 500 mV vs. NHE and (A2) Contact mode AFM image of Ir(111) obtained in air (1000 nm x 1000 nm, h_{max} = 5 nm) image of Ir(111) (B1) STM image of Ir(100) in air (500 nm x 500 nm, h_{max} = nm) and (B2) CVs of Ir(111) and Ir(100) in 1 M $HClO_4$ and 0.01 M $HClO_4$ obtained with a scan rate of 100 mVs^{-1} and 50 mVs^{-1}. Imaging conditions: STM: I_T = 1 nA, U_{bias} = 100 mV, SECPM: ΔU = 5 mV.

Both surfaces were electrochemically investigated in H_2SO_4 as well as $HClO_4$ in concentrations ranging from 0.01 M to 1 M. The CV in Figure 4.4 B2 shows Ir(111) in 1 M $HClO_4$ in black and Ir(100) in 0.01 M $HClO_4$ in red recorded with a scan rate of 100 mVs^{-1} and 50 mVs^{-1}. The hydrogen and the hydroxide adsorption regions are clearly visible for Ir(111) at typical potentials reported also in literature [177, 178]. Ir(100) shows a pronounced CV with all typical peaks reported in [164, 172, 177, 179]. The large peaks around 200 mV vs. NHE indicate the adsorption and desorption of hydrogen whereas the small peaks around 650 mV vs. NHE represent the adsorption and desorption of OH^- groups. Comparing this to the CV in [179] the very small prepeaks in the OH^- regime could not be observed in the present study. All peaks are not as sharp as they have been observed for perfect single crystals. The broadening of the peaks is an indicator for a smaller terrace width and a higher defect density compared to single crystals. Nevertheless, the electrochemical data are very sensitive to structural properties and still clearly show the characteristic properties of the Ir(111) and Ir(100) surfaces.

Single crystalline surfaces were investigated with three SPM techniques: STM, SECPM and AFM. In addition, cyclic voltammetry was used to investigate the quality and suitability of these surfaces for investigations in electrochemistry. Au(111) and HOPG were used as standard supports for calibration and well investigated surfaces in electrochemical environment. It was shown that the SECPM technique based on the potential difference between two electrodes in electrolytes has a resolution which is comparable to STM in air and under electrochemical conditions on the investigated surfaces. All presented heteroepitaxially grown Ru, Rh and Ir single crystalline surfaces show high quality obtained from the SPM and electrochemical investigations which is similar to bulk single crystals. Although some defects due to preparation were found on the surfaces, the quality represents a high standard and they are suitable for a wide range of investigations under electrochemical conditions.

4.2 Nanostructured Model Surfaces

4.2.1 Pulse Deposition Techniques

As described in Chapter 3, the deposition of Pd as well as Pt on Au(111) was mainly preformed using potentiostatic pulses. An advanced technique was used in later experiments creating Pt nanostructures applying a short nucleation pulse to create nucleation centers and afterwards a growth pulse to vary the amount of deposited metal. The duration of the nucleation pulse was 100μs and the final potential was 410mV vs. NHE. During the short pulse length a high time resolution of about 1μs was applied to obtain data with high time resolution. Figure 4.5 shows a current time transient and the charge time transient. From this follows that the resulting charge was 2.2μC for each experiment and varied only in the range of +/- 0.2μC.

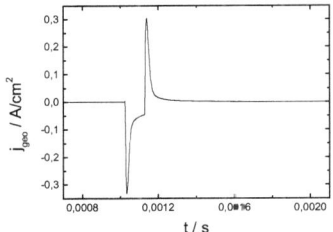
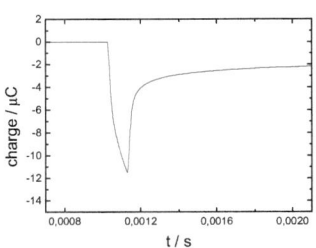

Figure 4. 5 Nucleation pulse for Pt deposition: left: current vs. time, right: charge vs. time.

For a further growing of particles on the support, a small overpotential between 10 and 120mV vs. equilibrium potential was applied. The standard potentials for the different complexes are listed in [312] An example is given in Figure 4.6 where the upper part the current time transient and on lower part the charge time transient is shown. The deposition current is constant after the first negative peak which indicates a diffusion controlled growth which is also seen in the linear behavior of the charge vs. time.

Several samples were produced by applying only a single potential pulse which is comparable to only applying a growth pulse with different overpotentials and different times.

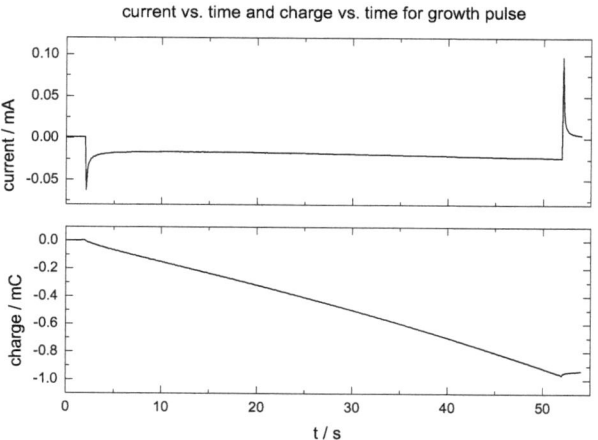

Figure 4. 6 Growth pulse for Pt deposition, top: current vs. time, bottom: charge vs. time.

4.2.2 Pd on Au(111)

In order to investigate nanostructured bimetallic surfaces a potential pulse deposition approach was used as described above. The current transient curve was used in order to determine the deposited amount of Pd by integration the current versus time. A full monolayer of Pd a charge is equal to $420\mu Ccm^{-2}$. A second approach to determine the amount of Pd is based on the charge transfer due to a hydrogen underpotential deposition in a 0.1M $HClO_4$ solution on the Pd-sites which is applicable for submonolayers and monolayers where no absorption into the Pd lattice occurs. The hydrogen adsorption method becomes less precise for small submonolayers, especially less than about 0.1 ML due to the very small hydrogen adsorption peaks in CVs. In parallel *in situ* EC-STM experiments were performed to image submonolayers of Pd which was deposited by potential pulse methods under the same conditions as in the standard glass cells. Here, the first monolayer Pd on Au(111) was completed before the second ML growths. Continuous Pd deposition in the multilayer deposition leads to three dimensional growth mechanism on Au(111).

Figure 4.7 shows different Pd/Au(111) surfaces with monitored wit STM under air. Pd was deposited from 0.1M HClO$_4$ + K$_2$PdCl$_4$. This characterization was performed in air after electrochemical characterization in perchloric acid or sulfuric acid. The amount of deposited Pd on the surface was determined by software and compared to the results obtained by electrochemical methods such as hydrogen adsorption and deposition charge for less than one monolayer which is in good agreement. Otherwise only the deposition charge was used to determine the amount of Pd.

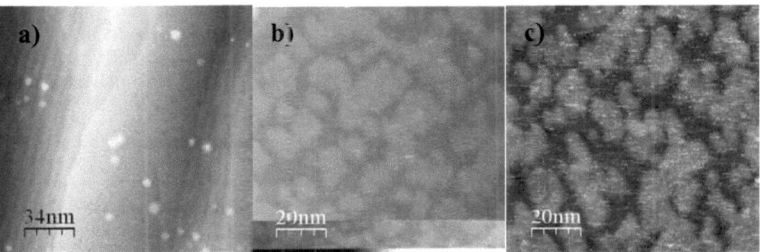

Figure 4. 7: Pd/Au(111) samples with different amount of deposited Pd and different resolutions; U$_{bias}$ = 100mV, I$_{tun}$ = 1nA, (a) 0.04ML , (b) 0.55ML, (c) 0.83ML of Pd on Au(111).

In addition, *in-situ* EC-STM experiments were performed to investigate the growth of Pd on Au(111) in 0.1M HSO$_4$ + 0.1mM PdSO$_4$. A small overpotential of 30mV for the palladium growth was applied and the increase of the amount of deposited Pd was recorded with the EC-STM (see Figure 4.8). A two dimensional growth for the first monolayer is visible if the potential of the WE is set to 700mV vs. NHE. Each picture was recorded in about 5min. It turns out that the growing behavior is according to Stranski-Krastanov mechanism explained in Chapter 2.

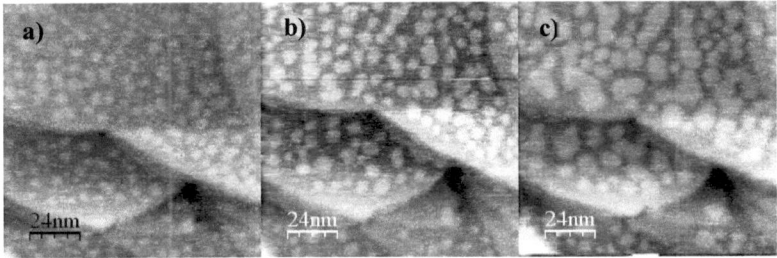

Figure 4. 8 In-situ Pd growth on Au(111) in 0.1M HSO$_4$ + 0.1mM PdSO$_4$; U$_{We}$ = 700mV *vs.* NHE; U$_{tip}$ = 900mV *vs.* NHE, I$_{tun}$ = 1nA, timescale from image (a) to (b) and image (b) to (c) about 5min.

4.2.3 Pt on Au(111)

Pt was deposited on Au(111) as described in Chapter 3 and 4. Pt was deposited from different electrolytes containing Pt^{2+} and Pt^{4+} ions. After depositing Pt on Au(111) the Pt coverage was investigated with STM for determining the amount of noble metal and the morphology of the particle size and distribution.

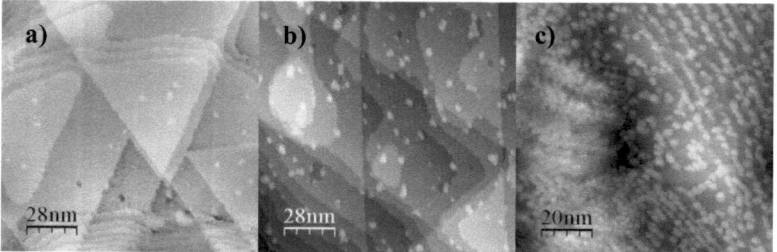

Figure 4. 9 STM image of different Au(111) electrode surfaces nanostructured with Pt; Different STM images recorded in air with U_{bias} = 100mV and I = 1nA. The amount of Pt on Au(111) corresponds to (a) 0.025ML, (b) 0.067ML, (e) 0.76ML of Pt.

In Figure 4.9 typical EC-STM images with different amounts of Pt deposited on the Au(111) surface are shown. The deposited amount of Pt was identified with WSxM software and recalculated to corresponding coverage.

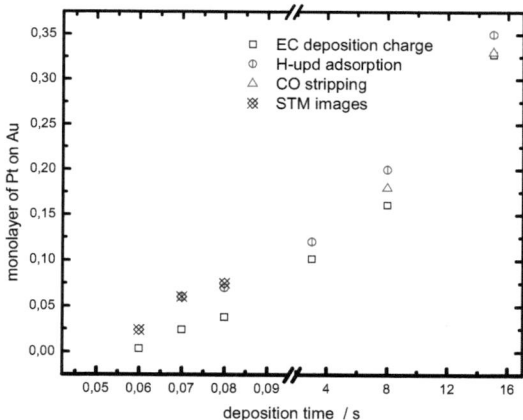

Figure 4. 10 Comparison of different methods which were applied in order to determine the amount of deposited Pt from 1M $HClO_4$ + 0.5mM K_2PtCl_6. The monolayer equivalent was calculated by electrochemical deposition (squares), STM images (open squares with cross), hydrogen adsorption (circles with line) and CO stripping experiments (triangles). The data is plotted versus the pulse length of the deposition pulses used for the nanostructuring of the Au(111) surfaces.

Different methods were used in this thesis to determine the amount of deposited Pt. Electrochemical deposition charge, hydrogen adsorption; CO stripping as well as STM was utilized in order to determine the coverage with Pt (see Figure 4.10). For very low coverages the amount of Pt was only derived from the Faradaic charge occurring within the electrochemical deposition processes.

As described in Chapter 3 the monolayer equivalents were derived by integration of current transients for all electrochemical experiments. For more than 10% of a monolayer the electrochemical results of the different methods are in good agreement. Also STM pictures were taken in order to evaluate the monolayer equivalent of the deposited Pt for less than 0.1ML. For all experiments the charge derived from the electrochemical deposition procedure has the lowest value evaluated. Both the values of the hydrogen adsorption charge and the coverage derived from EC-STM measurements exhibit higher values for the amount of Pt on the surfaces. This is most likely due to a spontaneous deposition of Pt after the pulse deposition process. At this time, the substrate is rinsed with water and for a few seconds in contact with the Pt-solution without potential control. In order to create single platinum nanoclusters, gold tips were loaded with platinum. Therefore a solution containing 1M $HClO_4$ + 0.5mM K_2PtCl_6 was used. Under potential control at 900mV vs. NHE the tip was dipped into the electrolyte and Pt was deposited for 60 seconds. The SEM images in Figure 4.11 show a bare Au tip and an Au tip was Pt deposits. The platinum loading is clearly visible in the SEM micrograph with islands of around 50nm in diameter

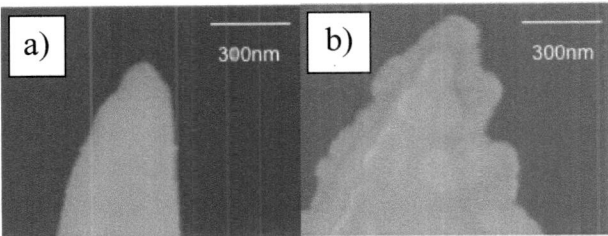

Figure 4. 11 a) Au tip, b) Pt deposited on a Au tip.

Figure 4.12 shows a Pt particle on Au(111) with different heights, where the difference between one and two monolayers can be seen. The one monolayer part as well as a two monolayer high part is marked and illustrated in the profile plot. The reconstruction of the Au(111) support is also clearly visible. Various particle heights were fabricated by different displacements of the STM tip in z-direction.

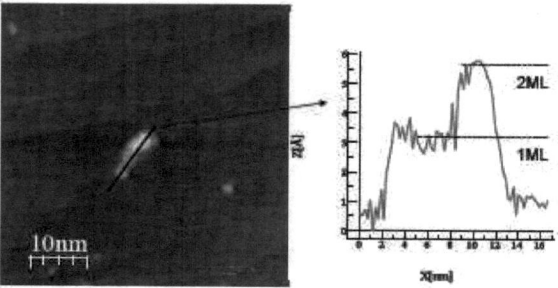

Figure 4. 12 Particle created in 0.1M HClO$_4$, the difference between two monolayers is visible, U$_{We}$ = 0mV vs. NHE, U$_{tip}$ = 350mV vs. NHE, I$_{tun}$ = 1nA, Right; cross section of deposited particle.

4.2.4 Pt on Ru(0001)

First experiments on Pt deposition on Ru(0001) were conducted to investigate the suitability of the single crystalline substrates for electrochemical purposes. Therefore Pt deposition was performed in 1M HClO$_4$ and 0.5mM K$_2$PtCl$_6$ using cycling voltammetry. The quality of the surfaces was checked before in perchloric acid and was as good as the results in Chapter 4. Starting at the initial potential of 860mV vs. NHE and sweeping with 100mV/s to 460mV vs. NHE leads to a deposited charge of 25μC which is equal to 0.15ML on the exposed 0.2cm^2 area of Ru(0001) shown in Figure 4.13. According to hydrogen adsorption on the Pt surface a charge from the CV of 9μC was determined after correcting the Ru background for the Pt/Ru(0001) sample. The deposition was continued with cyclic voltammgrams to further grow Pt to multilayers. A constant initial potential of 860mV vs. NHE was used with a final potential of 0mV vs. NHE with scan rates of 100mV/s and 10mV/s. In both cases the cycles were repeated ten times and a diffusion controlled deposition was observed. The total amount of Pt could not be determined by integration of the deposition cycles due to large overlapping with Ru features and hydrogen adsorption on already deposited Pt.

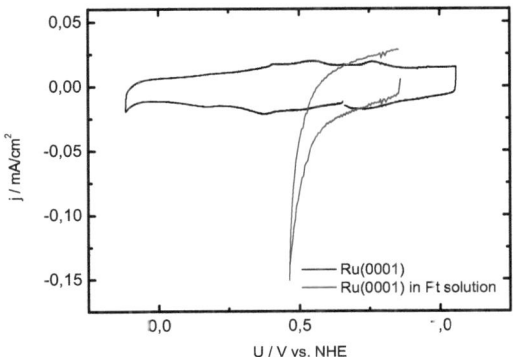

Figure 4. 13 CV on Ru(0001) in 1M HClO₄ with and without Pt.

Figure 4.14 shows three different Pt/Ru(0001) surfaces. The black line is without Pt, the thin line with 0.15ML Pt on Ru(0001) and the gray line with multilayers of Pt on Ru. The surfaces for the multilayer determination via adsorbed hydrogen lead to a charge of 80µC which can be calculated to an area of 0.38cm^2 assuming a charge of 210µC per cm^2. This also shows a surface roughness of about 1.9 compared to the smooth Ru(0001) support with a geometrical area of 0.2cm^2. At a given current density the overpotential for hydrogen evolution differs of more than 200mV comparing bare Ru(0001) with mulitlayers of Pt on Ru(0001).

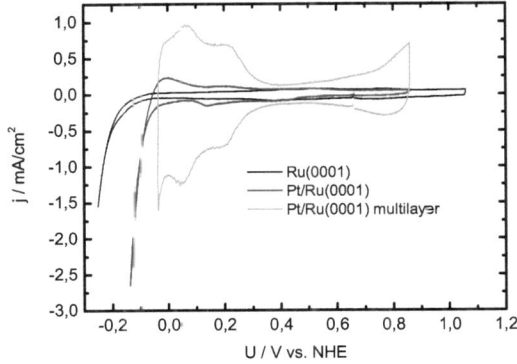

Figure 4. 14 CV of Ru(0001) and Pt decorated Ru(0001) between hydrogen and oxide formation of the surface in 1M HClO₄.

In order to investigate the multilayer deposition and to obtain morphology of the surface STM was performed on these Pt/Ru(0001) samples and hydrogen adsorption to determine the Pt surface. The multilayer structure of Pt on Ru is shown with STM in Figure 4.15 on the left side and compared to the bare Ru(0001) on the right side. As it can be clearly seen the Pt deposits are on the Ru and roughened the whole surface via a three-dimensional growth.

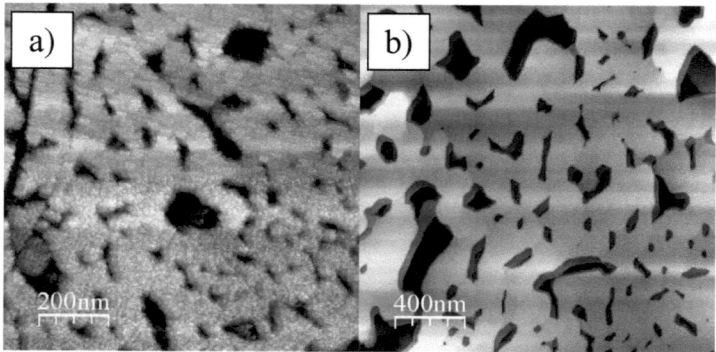

Figure 4. 15 a) Pt decorated Ru(0001), scan area 1μm^2, image surface area 1.47μm^2; b) Ru(0001) scan area 4μm^2, image surface area 4.04μm^2.

Calculating the image surface area with evaluation software (Nanoscope 7.30) the smooth Ru shows an area of 4.04μm^2 for 2μm x 2μm scan resulting in a surface area difference of 0.9%. The Pt decorated Ru surfaces shows an image surface area of 1.47μm^2 for 1μm x 1μm scan resulting in a 47% surface area difference compared to a flat undecorated surface. Although there is a difference in STM and hydrogen adsorption evaluation the results show the same tendency.

4.2.5 Cu-upd in Perchloric and Sulfuric Acid on Rh(111)

The quality of the Rh single crystalline surfaces was checked by cyclic voltammetry, as shown in Chapter 4. Noticeable differences between cyclic voltammetric profiles for Rh in HClO$_4$ and H$_2$SO$_4$ electrolytes were found and already described. Similar differences were also found on Au(111) and Au(100) already indicating the strong effect of adsorbing anions [313-315]. Hydrogen ad/desorption charge for Rh(111) is 256μC/cm^2 [316] for one full monolayer and was used to determine the surface area which was in good agreement with the geometric surface area.

The influence of anion coadsorption during the Cu upd on the Rh(111) electrode can be elucidated by comparing the data of Figure 4.16 and Figure 4.17 obtained in perchloric and sulfuric acid, respectively. Figure 4.16 shows the Rh(111) surface in pure perchloric acid and Cu ion containing perchloric acid where the upd and the Cu bulk deposition and dissolution is shown. It is well pronounced that Cu deposition and Cu dissolution occur in the dotted lines with well pronounced dissolution peaks at 510mV/490mV vs. NHE and 400mV/370mV vs. NHE for scan rates of 10mV/s and 2mV/s in 0.1M HClO$_4$. This results in a underpotential peak separation of 110mV and 120mV which is in line with recent literature [317] although the absolute values of the peak potentials differ due to the reference to RHE. In order to estimate the coverage of the underpotential deposited Cu on Rh(111) the upd peaks in the CV where evaluated. First results indicate that the Cu monolayer is not completely closed with a coverage of 0.5ML in perchloric acid assuming for a two electron process $512\mu C/cm^2$ according to hydrogen adsorption as one electron process. Below 340mV vs. NHE a continuous Cu deposition occurs overlapped by typical Rh(111) features and obviously diffusion controlled as indicated in the constant negative current until hydrogen evolution. The dissolution of Cu is fast and therefore independent of the sweep rate which is indicated with the sharp dissolution peak with roughly the same area. Prepeaks indicating a deposition on defects or different oriented surfaces where not determined which approves a homogenous Rh(111) electrode.

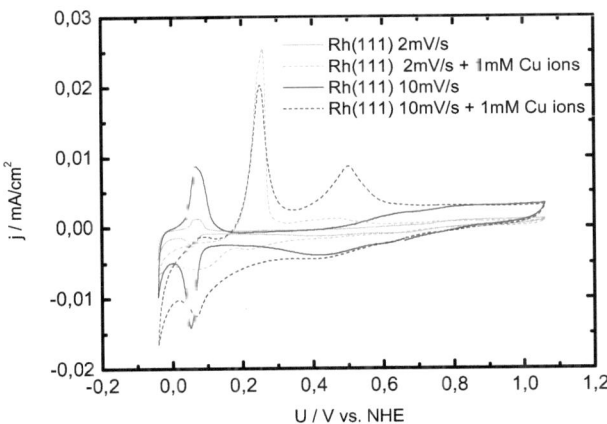

Figure 4. 16 Cu-upd on Rh(111), 0.1M $HClO_4$ + 1mM $CuClO_4$, scan rates 2mV/s and 10mV/s.

Figure 4.17 shows CVs in sulfuric acid with and without Cu^{2+} ions at different sweep rates. Although 100mV/s is quite fast to investigate the deposition of Cu as slow process all features are well pronounced. All details were also found at sweep rates of 10mV/s. The upd peaks are at 530mV/520mV vs. NHE for the Cu upd dissolution in 1M H_2SO_4. The Cu upd deposition can be roughly determined for 10mV/s sweep rate to 300mV vs. NHE which leads to a ΔU of 220mV. A comparison with [317] shows that the peak potentials for dissolution are equal but the potential difference ΔU is 30mV larger compared to the value in the reference. An evaluation of the charge which is transferred during the Cu upd monolayer deposition and dissolution shows that the upd layer is not completely closed and calculated to approx. 0.8ML. This result is compared to 0.9ML for an upd layer in [317] smaller which may be due to the faster sweep rates compared to those in the reference. At potentials lower than 340mV Cu is continuously deposited onto the Rh(111) leading to a diffusion controlled deposition with overlapping Rh features (seen in Figure 4.17).

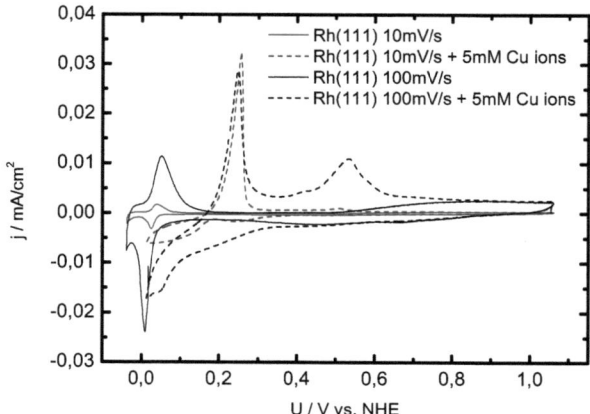

Figure 4. 17 Cu-upd on Rh(111), 1M H_2SO_4 + 5mM $CuSO_4$, scan rates 10mV/s and 100mV/s.

Comparing the Cu deposition on Rh(111) in different electrolytes leads to the statement that the anion effect is more pronounced on Rh(111) for sulfuric acid as compared to perchloric acid. Adsorption of (bi)sulfate [166] shows a clear influence on the position of the UPD peaks which can be clearly seen in the shifted dissolution peak in perchloric electrolyte to more positive potentials when referring to RHE by adding 60mV regarding the 0.1M $HClO_4$ and 1M H_2SO_4 electrolytes. The equation $\Delta U = \alpha \Delta \phi$ [318] where ΔU denotes the experimentally observed upd shift, $\Delta \phi$ the differences in work function between deposits and supports and the constant $\alpha = 0.5V(eV)^{-1}$ can be used to compare the experimentally obtained values with literature ones. The work function of Rh(111) is 5.4eV and given in [319] Cu(111) has a work function 4.98eV which is given in [320]. For the Rh(111)/Cu system this results in a value of 210mV for the upd shift which is very close to the experimental value of 220mV in sulfuric acid and only slightly different from that obtained in perchloric acid.

Various deposition techniques were applied in order to create nanostructures on single crystal supports, ranging from single particles with the EC-STM to multilayer structures via electrochemical deposition. Pd and Pt deposition on Au(111) was the major task for further reactivity measurements. Pt deposition on Ru(0001) and Cu deposition on Rh(111) provide the utilization of the new support materials.

4.3 HER/HOR, ORR and MOR on Pd/Au(111) and Pt/Au(111)

4.3.1 HER/HOR on Pd/Au(111)

On Pd modified Au(111) surfaces reactivity measurements regarding hydrogen reactions were carried out in H_2 saturated 0.1M $HClO_4$ solution using potentiostatic pulses at various overpotentials. The current transients for the HOR were evaluated using j vs. $t^{1/2}$ plots in order to separate the kinetic current from the mass transport current. The obtained kinetic currents were used to construct Tafel plots. Tafel plots for different amounts of Pd deposited on Au(111) electrode surfaces were evaluated with respect to the geometrical area of the substrate and with respect to the Pd area. The latter assumes that only Pd is active in the reaction. The current with respect to the geometric area of the Au(111) surface increases with decreasing amount of Pd from 0.74ML to 0.1 ML by about one order of magnitude with a maximum value of 1mA/cm^2 for an overpotential of 350mV. The Tafel plot of the specific current density, where the current is normalized to the surface of deposited Pd, is shown in Figure 4.18. The data indicates that the specific reactivity of Pd increases by about two orders of magnitude with decreasing amount of Pd from 0.74ML to 0.1ML. Hence, current densities of up to 10mA/cm^2 per Pd area are observed.

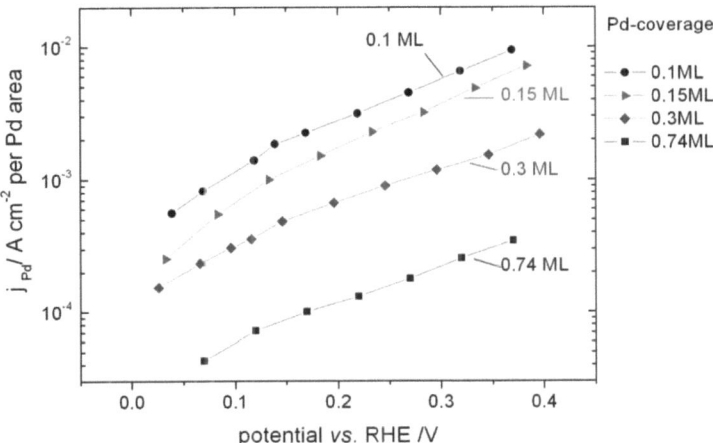

Figure 4. 18 Tafel plot of HOR, specific current density is plotted vs. potential for Pd submonolayers and monolayers.

4.3.2 HER/HOR on Pt/Au(111)

The Pt/Au(111) nanostructured surfaces were prepared as described above and coverages were varied in the range of 0.0005ML to 7.8ML Pt on Au(111). Potential step experiments as well as micropolarization curves in the range of +/- 10mV near the equilibrium potential were used to investigate the reactivity reflected in the exchange current density. In Figure 4.19 the exchange current density is plotted versus the coverage of Pt on Au(111). Both measurement techniques, namely potential pulse methods (circles in Figure 4.19) and micropolarization curves (squares in Figure 4.19) show the same exchange current density j_0 versus coverage of platinum on Au(111). The x- and the y-axes are in logarithmic scale to deal with the large range variations in coverage and in exchange current density. As reference a Pt(111) single crystal was investigated and showed a behavior which is in line with literature [96-98, 100, 133]. For Pt monolayers on Au(111) the reactivity is already larger as compared to Pt(111). A slight increase in exchange current density with decreasing monolayer thickness from 7.8ML to 0.1ML of noble metal is shown. The increase in exchange current density rapidly increases, however, at coverages lower than about 10% of a monolayer. An increase of more than two orders of magnitude for the hydrogen reactions was found in this region of the low coverages. Especially the values for coverages below 10% were measured with two different techniques, namely potentiostatic pulse and micropolarization curves and shows good agreement.

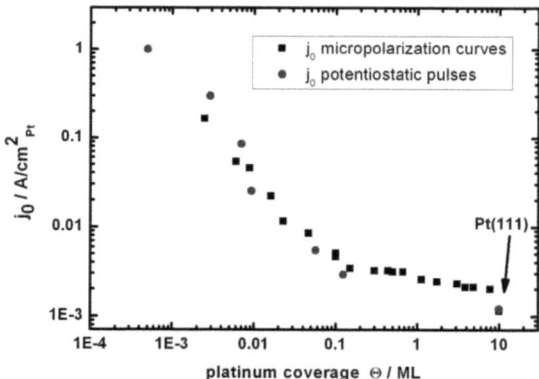

Figure 4. 19 Exchange current density for hydrogen reactions *vs.* Pt coverage on Au(111); values were obtained from potentiostatic pulses (circles) and micropolarization curves (squares).

Results with single Pt particles show a similar dependence of the exchange current density when plotting versus the particle height, given in atomic layers (AL) shown in Figure 4.20 [321]. Particles with a height between one atomic layer and five atomic layers were investigated and the reactivity measurements indicate a strong increase with decreasing thickness of the particle from five AL down to only one AL height.

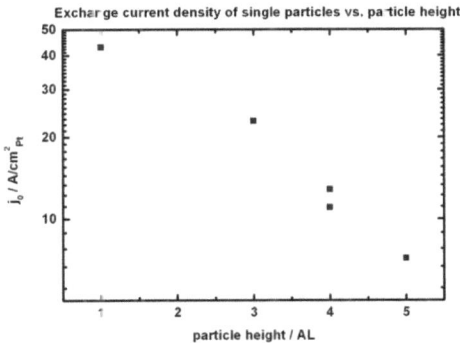

Figure 4. 20 Exchange current density vs. particle height of single Pt particles on Au(111).

The exchange current densities were determined as described above and values between 3A/cm^2 and more than 40A/cm^2 were found. The linear behavior in a logarithmic plot is indicating an exponential correlation between particle height and exchange current density (see Figure 4.20). according to the Pd/Au(111) system [322].

4.3.3 ORR on Pd/Au(111)

The deposition and characterization of Pd mono- and submonolayers were carried out as described before. After purging the electrolyte with oxygen the initial potential was set to 820mV vs. NHE and potential pulses with varying overpotential for the Pd/Au(111) system. For the Pd/Au(111) system the current densities were derived from potentiostatic pulse studies and the extrapolated kinetic current densities are summarized in a Tafel plot (see Figure 4.21). Au(111) as support shows a very low activity towards the ORR, nevertheless the currents for the coverages smaller than 0.33ML were corrected for the gold offset. An increasing activity with increasing amount of Pd can be observed starting from 0.11ML to 2.1ML. Furthermore, the onset of the ORR is shifted to more negative potentials for lower amounts of noble metal.

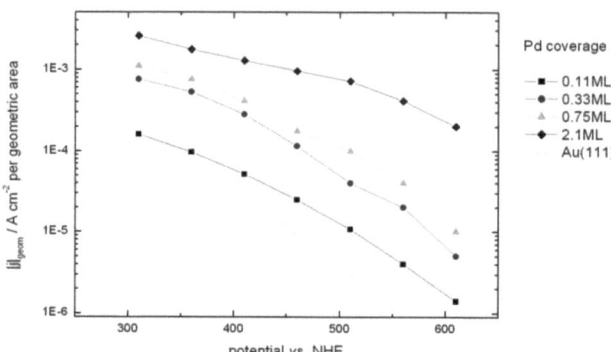

Figure 4. 21 Tafel plot of ORR; specific current density is plotted vs. potential for Pd submonolayers and monolayers.

Relating the current densities to the active surface area (shown in Figure 4.22) the specific activity is constant or is slightly increased for larger amounts of palladium for 410mV and 510mV vs. NHE. Diffusion effects during the ORR may influence the results in the potential range smaller than 310mV vs. NHE.

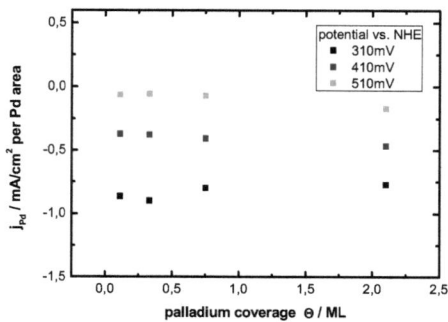

Figure 4. 22: Specific current density vs. palladium coverage for Pd/Au(111) surfaces for ORR.

4.3.4 ORR on Pt/Au(111)

The deposition and characterization experiments of Pt submonolayers were carried out as described before. After purging the electrolyte with oxygen, the initial potential was set to 820mV *vs*. NHE and cyclic voltammograms for the Pt/Au(111) system were performed between 820mV *vs*. NHE and 250mV *vs*. NHE. The obtained results for the Pt/Au(111) system were summarized in a Tafel plot. For a better comparison with results for the hydrogen and methanol reaction the current density for a given potential was plotted versus the coverage of Pt on Au(111), shown in Figure 4.23. All current densities are referred to the amount of platinum to compare specific current densities of the active Pt catalyst. Results obtained on a Pt(111) single crystal surface are added for comparison. It can be seen that the reactivity of the Pt(111) surface is similar to the reactivity of 7.8ML Pt on Au(111) and shows the highest activity in all cases.

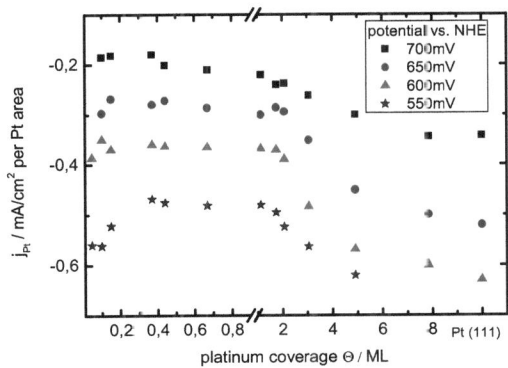

Figure 4. 23 Specific current density of Pt for oxygen reduction versus platinum coverage on Au(111) for different potentials. The values were obtained from CVs in O_2 saturated 1M $HClO_4$.

Lower coverages show an increasing current density with increasing Pt coverage for a potential of 700mV vs. NHE. At lower potentials a similar behavior for coverages larger than 30% is shown. A slightly different behavior for the reduction of oxygen was found for coverages smaller than 30% for the potentials for 550mV vs. NHE. The current densities are somewhat higher compared to larger coverages. This behavior was not investigated in detail but diffusion effects may play a key role when lowering the coverage and thus providing island catalysts with a changed transport mechanism of educts and products in contrast to extended catalyst surfaces. Summarized, the Pt/Au(111) shows an activity increasing with increasing amount of deposited Pt regarding ORR for low overpotentials. Higher overpotentials result in two different behaviors for the activity. For coverages larger than approx. 30% the activity is increasing with increasing amount of Pt. Coverages smaller than approx. 30% show an increasing behavior but with decreasing amount of Pt for high overpotentials. This results in a current density minimum at about one ML.

4.3.5 MOR on Pd/Au(111) and Pt/Au(111)

Deposition and characterization Pt submonolayers and monolayers were carried out as described before. The methanol oxidation reaction was measured in 1M $HClO_4$ + 1M CH_3OH Ar purged electrolyte using cyclic voltammograms with a scan speed of 100mV/s. A typical double peak structure was obtained. The maximum current and the potentials of the forward and the reverse peak were determined. The results of the evaluation of the maximum current are summarized in Figure 4.24 top, showing a strong dependence of the Pt coverage on the activity towards methanol oxidation.

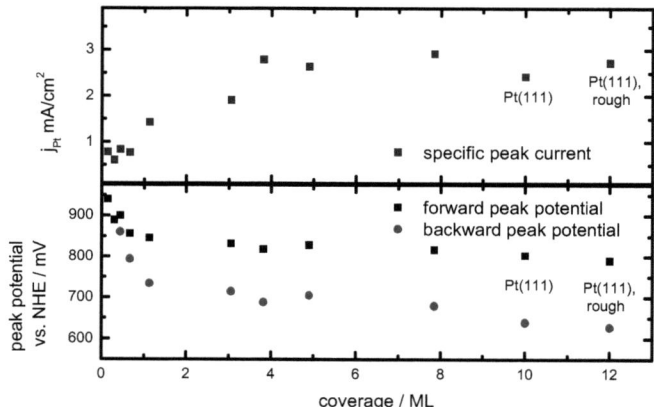

Figure 4. 24: top: Specific peak current density of Pt for methanol oxidation versus platinum coverage on Au(111); down: Peak potential of methanol oxidation for forward and reverse cycle versus platinum coverage on Au(111). The values were obtained from CVs in Ar saturated 1M CH_3OH + 1M $HClO_4$.

For very low coverages in the range smaller than 0.15 monolayers no reliable currents could be determined from the CVs. It was found that an increase of the maximum current with increasing Pt coverage on Au(111) up to approx. 3ML occurs. Increasing the Pt coverage results in a constant maximum current, which is in line with single crystal surfaces. For comparison a Pt(111) well oriented and a rough Pt(111) single crystal is shown, indicating that the Pt(111) is less reactive compared to more polycrystalline surface of the roughened Pt(111). The measured maximum current density of 3mA/cm^2 is in line with literature [133, 255].

For comparison of the peak potentials the forward and the reverse peak potential of the methanol oxidation was evaluated and plotted versus the Pt coverage on Au(111) (see Figure 4.24 down). Due to the high scan velocity of 100mV/s the forward and the reverse peak potentials are separated whereas the forward potential has the higher value. The separation of the peak potentials depends on the Pt coverage and is at least 100mV. The rough Pt(111) has the lowest onset potential slightly lower compared to the well prepared Pt(111). Between 7.8ML and 1ML of Pt on Au(111) the peak

potential increases with decreasing in a moderate way. A strong increase of the peak potential is observed below one ML of Pt on Au(111) resulting in a total shift of at least 150mV compared to the submonolayers regime with the roughed Pt(111).
The methanol oxidation was also investigated on the model surface Pd/Au(111). Even monolayers and multilayers were inactive towards the methanol oxidation at the standard experimental conditions. The investigation of Pt towards the MOR was therefore not investigated in detail.

4.4 Local pH Sensor, Pd/H Electrode

It was shown in the past, that the EC-STM tip can be used as a local sensor to measure the reactivity of single Pd [19, 21] or Pt nanoparticles via the tip. That method detects at the tip the reaction products of a catalytic active particle on a surface. In the cited cases a Pd or Pt particle deposited on Au(111) evolves hydrogen which is then oxidized at the tip and induces a current flow.
This idea was transferred to locally determine also the proton concentration which varies for several reactions generating or consuming protons. The SECPM offers this possibility due to the potentiometric measurement technique using the tip as a local potential sensor which is held at open circuit potential (OCP) since a change in proton concentration changes the local pH a proton sensitive tip senses a potential shift of 59 mV per pH value. Thus, hydrogen evolution or oxidation can be directly detected at the SECPM tip by a shift of its OCP. Therefore pH sensitive tips such as Pt and Pd where used to determine their behavior on the required nanometer length scale. Due to their ability to store hydrogen, Pd tips were favored and investigated in detail.
Polycrystalline palladium was loaded with hydrogen by galvanostatic pulses to form palladium hydride electrodes in an electrochemical glass cell according to [178]. A hydrogen ab/adsorption and desorption process is shown in Figure 4.25. The applied current is shown in black with the axis on the right side and the measured potential is shown in thin black with the axis on the left side. Applying a negative current in the neutral 1M $NaClO_4$ electrolyte leads to a genitive potential at -1.3V vs. NHE where hydrogen evolution and hydrogen absorption occur in parallel. At 1100s the OCP of the tip is measured for about 400s. The OCP is not stable and increasing with time. The short positive current pulse for duration of 100s causes a high OCP which stabilizes after several ten seconds. Repeating the procedure with a negative current pulse leads again to unstable OCP, seen in the time range between 2150s and 2350s. The last positive current pulse which was applied with two different current steps shows the same beginning of the transient as compared to the first positive current pulse.

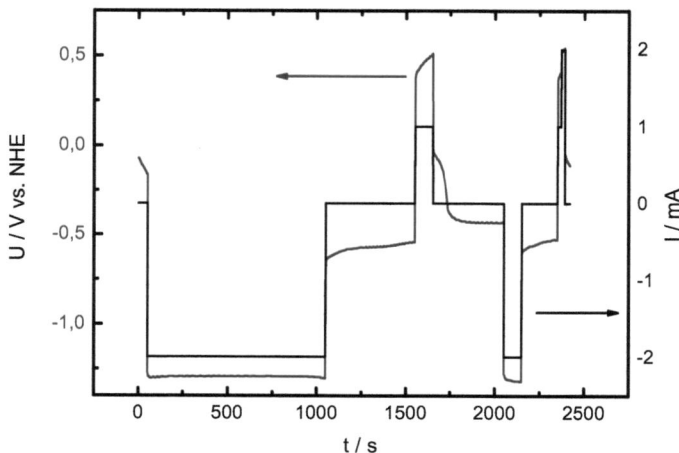

Figure 4. 25: Galvanostatic pulses on Pd electrode; Potential response due to the galvanostatic hydrogen ad- and absorption, hydrogen desorption and OCP measurements.

A closer look into the steps forming a stable Pd/H electrode can be seen in Figure 4.26. The initial potential drop includes double layer charging, hydrogen adsorption and Pd hydride formation. The completion of the hydride phase is at the end of the first potential plateau. The further potential decay and thus the increase in overpotential are attributed to hydrogen evolution and further hydrogen absorption into the palladium. These results are comparable to potential step experiments for form Pd/H and detecting current transients in [323]. The OCP measurement after the hydrogen loading starting at 610s already shows the very high stability of the produced electrode.

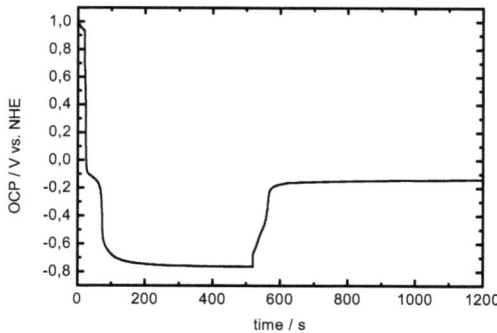

Figure 4. 26: Formation of Pd/H electrode and subsequent OCP measurement. The applied negative current for hydrogen loading was -500µA for 500s.

As already shown in Figure 4.25 also positive galvanostatic pulses were applied to obtain stable OCP values of the Pd/H electrode. Based in the work of Flanagan and Lewis [289, 324] the electrochemically loading of palladium electrodes can lead to a higher concentration of surface hydrogen than hydrogen in the bulk for the hydrogen/palladium ratio larger compared to 0.4. Therefore, a positive pulse after hydrogen loading was applied to the Pd/H electrode in order to remove excess surface hydrogen and relieve concentration gradients [291].

Titration experiments in order to investigate the pH sensitivity of the Pd/H electrodes were performed in standard glass cells. The Pd/H electrode was prepared in 1M $HClO_4$ Ar purged electrolyte as described above. Starting at a pH value of approximately 5.8 in 1 M $NaClO_4$ (80 ml) electrolyte the proton concentration was increased by adding stepwise $HClO_4$ in different concentrations ranging from 0.1M to concentrated $HClO_4$. During the whole experiment the OCP of the Pd/H electrode was recorded. The potential transient of the Pd/H electrode versus time is shown in Figure 4.27. Due to the strong dependence of the OCP to the proton concentration und the low concentration of protons in a pH 5.8 electrolyte only 10µl of 0.1M $HClO_4$ was added in the first step. This already leads to a potential step of about 100mV. Continuous adding of perchloric acid in different amounts and concentrations is shown of Figure 4.27. The OCP shifts to more positive values. At the end of the experiment the pH of the electrolyte was 0.36 determined with a pH meter. From the amount and concentration of added $HClO_4$ the pH value of the electrolyte can be calculated after each step.

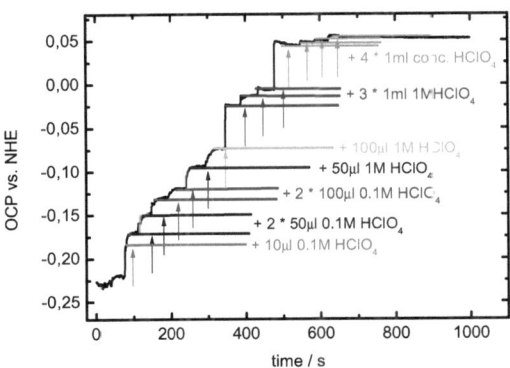

Figure 4. 27: Titration curve of Pd/H electrode. OCP vs. time

This experiment was also done adding alkaline NaOH electrolyte in different amounts and different concentrations. A shift of the OCP towards more negative potentials was recorded and is in line with the results found in acidic environment.

In order to apply all the results of the Pd/H electrode the surface of the Pd has to be minimized for use in an EC-STM or SECPM. This was achieved by insolating the etched Pd electrodes with Apiezon wax as described in Chapter 3. The surface of the Pd tips was determined with different techniques such as double layer charging method, Fe^{2+}/Fe^{3+} redox couple method and hydrogen adsorption. All methods are used in a limit range of their field of application. Nevertheless the results of the determined surface areas are for each experiment in the same order or magnitude which implies that the radii of the free tip apex only varies of a factor three assuming a hemispherical shape.

The relation between the OCP of an insulated Pd/H tip and the pH value is shown in Figure 4.28. The black curve is a guide for an eye with a slop of 59mV/decade as theoretical value [62, 323]. Although the measured data are not on a perfect line the slope of 60mV/decade is very close to the theoretical value. The intersection point with the y-axis is at 60mV vs. NHE which is also very close to the literature value of 50mV vs. NHE [291, 293].

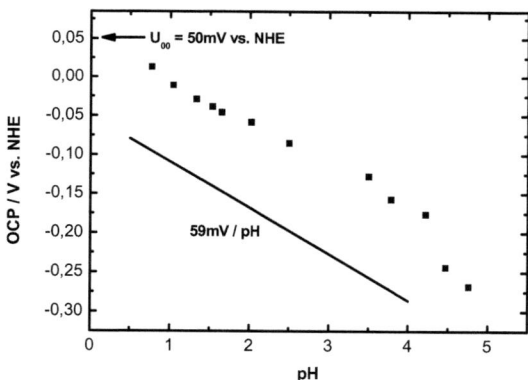

Figure 4. 28: pH sensitivity of the Pd/H electrode OCP.

The stability of the OCP for the nano pH sensitive electrode was a challenge to investigate. As shown above large Pd/H electrodes show good stability and behave as already described in literature. Figure 4.29 pointed out that also an insulated with hydrogen loaded Pd STM tip is comparable to a large electrode towards its OCP.

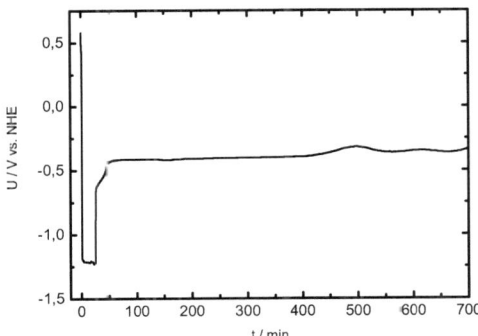

Figure 4. 29 OCP vs. time for a hydrogen loaded Pd tip.

The potential starts with a positive value which is undefined. A negative current of -10µA leads to potential drop due to forming a Pd hydride phase, hydrogen evolution and hydrogen absorption. A subsequent measurement of the CCP shows after a short relaxation time a stable value for about 4 hours. After this time a slight increase in potential is seen which then requires a new loading with hydrogen showing a stable potential behavior and thus suitable to investigate pH changes over a long time period.

In order to expand the local technique using the tip of an STM as current sensor in electrochemical environment, a local potentiometric tip was developed. Hydrogen loaded palladium serves as a stable and precise pH meter due to the fixed potential of the electrode determined by the absorbed hydrogen and the concentration of the protons in the surrounding solution. Galvanostatic pulse loading was successfully used to prepare these kinds of nanoelectrodes and titration experiments showed the pH – potential dependence as described in literature for larger hydrogen palladium systems. Potential stability is shown for more than four hours.

5 Discussion

The discussion is subdivided into three parts: fundamental electrochemical studies on a new kind of model surfaces for electrochemistry; local techniques based on EC-SPM in order to investigate the activity of single particles in electrochemical environment and the main part: support effects for hydrogen reactions, oxygen reduction and methanol oxidation.

5.1 New Supports in Electrochemical Environment

Single crystalline surfaces were investigated with three different SPM techniques; STM, SECPM and AFM, which are based on very different physical phenoma such as the quantum tunneling effect in STM, a measurement of the potential difference in SECPM and using force as governing factor in AFM. In addition, cyclic voltammetry was used in order to investigate the quality and suitability of these surfaces for investigations in electrochemistry.

The new SPM technique SECPM which is based on the potential difference between two electrodes in electrolytes was used for in-situ characterization of single crystal electrodes with a resolution which is comparable to STM and AFM. Surfaces such as Ru and HOPG imaged with SECPM show a higher image contrast on steps compared to Au, Rh and Ir. This irregularity at the step edges may arise due to different surfaces states. Potential spectroscopy as well as high resolution scanning with SECPM can provide new insights into surface science phenomena under electrochemical conditions in future.

All presented heteroepitaxially grown Ru, Rh and Ir single crystalline surfaces show large single crystalline regions almost as bulk single crystals. The preparation of these metal surfaces on 4 inch Si wafers [325] yield a large number of identical samples for single-use electrodes which is of special interest when decorating with foreign metals providing a high through put screening. All results were confirmed by SPM and electrochemical investigations and discussed with bulk single crystal surfaces results from literature. Although there are defects due to preparation the surfaces quality has very high standard which makes it suitable for a wide range of investigations under electrochemical conditions. Especially, if the support material is decorated with foreign metal in order to clarify the influence of the support. The nanostructred surfaces will provide new insights in fundamental understanding of electrocatalysis as in the case of the Pd/Au(111) and Pt/Au(111) investigated in detail in this thesis.

Rh(111) was chosen as support for principal electrochemical studies for potential induced metal deposition due to very recent results [317] regarding Ag and Cu deposition on Rh single crystal surfaces. The underpotential deposition of copper is one of the most prominent phenomena in electrochemical metal overlayers due to good control of deposition and dissolution of Cu. Cu-upd can be used to determine the electrochemical active surface are and the difference in work function between deposits and support [318, 326, 327].

As a model catalyst system Pt was deposited on Ru surfaces. The bare Ru shows an inactive electrochemical behavior with a high overpotential for the hydrogen evolution reaction compared to Pt/Ru surfaces. An amount of 0.15ML Pt does not show the well pronounced Pt features such as hydrogen adsorption and desorption. Mulitlayers of Pt on Ru show typical polycrystalline Pt CVs with well defined peaks in the hydrogen region. In contrast to Au(111) as support material the STM images and the electrochemical results do not indicate a layer by layer growth for the first monolayers in the case of Ru as support material. Support effects which are strongly dependent on the layer thickness and on an equal growth mechanism are a challenging task for Ru as substrate material. Nevertheless, the bare Ru(0001) shows single crystal behavior demonstrated with STM and electrochemical methods; a deposition of foreign metals was easily possible. Basic CVs were obtained und compared to results obtained by STM which fits well together. For the first results with respect to the hydrogen evolution one can state that larger amounts of Pt causes higher activity. A detailed investigation of different coverages and different growth procedures would clarify the impact of Ru as support in detail.

5.2 Local Approaches for Reactivity Measurements

5.2.1 Local Current Technique

The basic idea of local reactivity measurements is to use the EC-STM tip as a local sensor to measure currents. The reaction products of an active species on the working electrode are reconverted at the tip into a faradaic current. Here the hydrogen evolution reaction at a single particle was investigated by measuring the reverse hydrogen oxidation reaction at the STM tip. The collection efficiency of the STM tip is approximately 1. A sequence of steps has to be done in order to perform local reactivity measurements and a detailed analysis of the data is necessary.

Former results from Meier et al. [19, 21] already showed a strong dependency of the particles height on the catalytic activity for the system Pd/Au(111). Here, smaller Pt particles were generated and investigated ranging between one and five atomic layers in height. An increasing activity of almost one order of magnitude with decreasing particle height was found. The technique was able to monitor the electrocatalytic activity for these small particles on the nanometer scale and compare it to large nanostructured surfaces. As seen in Figure 5.4 single particles exhibit exchange current densities of more than 6A/cm^2 which are purely kinetic since hemispherical diffusion.

5.2.2 Local Potential Technique

Performing spatially resolved pH measurements at the nanoscale using electrochemical SPM techniques is of great interest in the research field of electrocatalysis [328] and corrosion [292]. Similar to the approach shown above where the EC-STM tip was used as a local current sensor first attempts towards a potentiometric SECPM tip as a local pH sensor were performed. Whereas the potentiometric properties are not thought to be size dependent, at least as long as the response of the surface surpasses that of the edges, i.e. with smaller electrodes imperfections will probably become critical and seriously affect the electrode potential [323]. Most potentiometric pH microsensors are covered with glass, liquid and polymeric membranes, or metal oxide films [329]. However, these materials are not suitable for the application as electrochemical SECPM tip. Therefore, a pH sensitive tip material such as platinum (Pt) or palladium (Pd) is required resulting in first steps towards a suitable sensor.

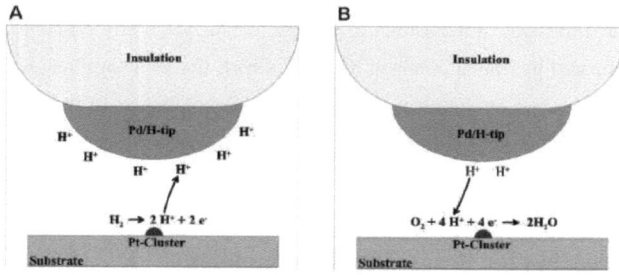

Figure 5. 1 Potential shift method: Schematics of the proton concentration measurement with a hydrogen loaded palladium SECPM tip (Pd/H tip) as a local pH sensor. A) Illustration of increasing proton concentration; B) Illustration of decreasing proton concentration.

The potential of a Pd/H electrode is dependent on the composition in the pure α and β phases as described in Chapter 2, but is independent of composition in the mixed $\alpha + \beta$ phase region. Therefore using a Pd/H SECPM tip electrode in the miscibility gap, a change in the proton concentration changes the equilibrium potential of the Pd/H electrode. The high impedance amplifier of the SECPM setup allows the measurement of the open circuit potential (OCP) at the SECPM tip. Therefore, the potential shift of the OCP caused by an increase or decrease in proton concentration within a catalytic reaction can be examined (schematically illustrated in Figure 5.1).

According to the Nernst equation, a change of the proton concentration by a decade results in a change of the measured OCP of the tip by 59 mV. This effect can be used to set up a method to directly determine the proton concentration at the SECPM tip.

All above mentioned phenomena were observed on polycrystalline Pd electrodes as well as on electrochemically etched and insulated Pd tips in Chapter 4. The hydrogen loading under galvanostatic control was successfully used to load and stabilize the potential of the electrodes. Also a change of the OCP using a hydrogen loaded Pd tip in the SECPM was observed varying the concentration of the electrolyte.

The basic principles of the local measurement due to proton concentration were elaborated with large as well as small Pd electrodes. Hydrogen palladium is a suitable system to apply in SECPM and very sensitive towards pH changes ranging from the acidic to alkaline solution. A combination with the above discussed local current measurements in future can provide a powerful tool for local investigations in the nanometer scale.

5.3 Support Effects

Support effects are the key topics of this thesis and will be separately discussed for the reactions but with the main focus on hydrogen reactions. Local as well as large nanostructured electrodes and their specific catalytic activity for hydrogen reactions is shown and discussed in detail. Pd and Pt depositied on Au(111) show increasing specific catalytic activity for decreasing amount of noble metal. In contrast to that, the geometric current densities show different behavior for Pd and Pt deposited on Au(111).

5.3.1 Hydrogen Reactions – Specific Reactivity

The hydrogen oxidation reaction and the hydrogen evolution reaction were investigated on Pd and Pt decorated Au(111) surfaces in hydrogen saturated electrolyte. Different electrochemical methods such as cyclic voltammetry and potential pulses were used in order to determine the catalytic activity which is reflected by the measured current. The reactivity of Au(111) surfaces covered with a large number of Pd and Pt particles, multilayers of Pd and Pt as well as single Pt particles deposited on Au(111) show an effect in that:

- The exchange current density of single Pt particles on Au(111) increases with decreasing particle height from five to one atomic layer by almost one order of magnitude;
- The absolute values for exchange current density of single Pt particles are in the range between 6 A/cm^2 and 50 A/cm^2 depending on particle height;
- Decreasing Pt coverage from 7.8 multilayers down to less than 1‰ of a monolayer on Au(111) leads to an increase of the exchange current density by more than three orders of magnitude;
- An increase of a factor of three in specific exchange current density is found comparing Pt(111) and one monolayer of Pt on Au(111);

- The specific exchange current density for Pt on Au(111) increases for coverages of less than 10% of Pt on Au(111) by more than two orders of magnitude;
- For coverages below 0.005ML Pt on Au(111) an exchange current density of up to 1 A/cm^2 can be observed;
- With decreasing coverage of Pd on Au(111) in the submonolayers regime even the geometric current density for Pd increases;
- The specific current density for Pd on Au(111) increases by two orders of magnitude in the submonolayers regime.

This strong increase in specific reactivity of Pd and Pt nanoislands on Au(111) will be discussed in the following:

1) The lattice of the Pd and Pt nanoislands experiences a strain due to the larger lattice constant of the Au(111) support, strain decreases on the layer thickness. Thus, the binding energy to the adsorbed hydrogen [16, 19, 33, 35, 322] is changed. A higher activity of a monolayer of Pt and Pd compared to multilayers on Au(111) is the result, because the strain of the adlayer disappears with increasing layer thickness. The same explanation can be applied for decreasing particle height for single Pt and Pd particles. In addition, also a modified binding energy to hydrogen for submonolayers and small clusters is reported [330] and investigated in collaboration with theoretical groups.

2) A spill over concept can be taken into account considering the binding of atomic hydrogen on Pt and Pd as well as on the Au support [15].

3) A high number of reactive steps and defects may cause an increased specific activity [6, 7] for large Pt and Pd nanostructured surfaces as well as single Pt and Pd particles on Au(111). The ratio of terrace atoms to edge atoms decreases with decreasing coverage of Pt and Pd on Au(111)

4) An enhanced mass transport for small particles [18, 214] due to hemispherical diffusion as compared to planar diffusion on extended surfaces may lead to higher current densities for very small coverages and for single Pt and Pd particles on an extended single crystal.

ad 1) Assuming a pseudomorphic overlayer on Au(111) the lattice of the Pd and Pt ad-layer on Au(111) is strained by approx. 4.8% (Pd) and 4% (Pt). According to the Nørskov model [30, 33, 35] a shift of the center of the d-band to higher energies results. This leads to a higher H-adsorption energy on Pd/Au(111) and Pt/Au(111) as compared to bulk Pd and Pt. Even a changed binding energy of hydrogen to different metals for submonolayers was reported [330] and can be applied for coverages smaller than a complete monolayer. Binding energies and reaction barriers of clusters with a few atoms and monoatomic rows of Pd were obtained in cooperation with theoretical groups. It was found that supported low coordinated Pd atoms embedded in the surfaces in different ways have an enhanced the activity as compared to bulk material which is shown in Figure 5.2 for the HOR.

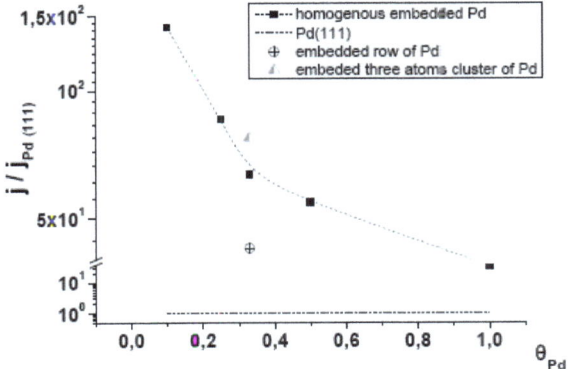

Figure 5. 2 Theoretical values of the exchange current for the HOR normalized to the surface of Pd(111) for different composition of palladium embedded in a surface of Au(111) calculated with the electrocatalysis theory. For comparison, the values obtained with a embedded row and a embedded three Pd atoms clusters in the Au(111) are included. The dotted black line indicates the value of 1 for the Pd(111) surface.

Santos and Schmickler [42, 331, 332] calculated potential energy surfaces of the bond length between two hydrogen atoms and determined the saddle point of the reaction pathway. These findings lead to the assumption that desorption of atomic hydrogen is also an important step. Especially a recent publication from Santos et al. [38] predicted the enhanced reactivity of a monolayer of Pt on Au(111) and Pt nanoparticles on Au(111) for hydrogen evolution which is in line with results reported here. Kinetic data for a comparable coverage of Pt on HOPG yield specific current densities lower by orders of magnitude, emphasizing the importance of the substrate [20]. This is also supported by measurements of Pd particles on Cu(111) [333] and Pt nanostructures on Cu(hkl) [334] where no enhancement of reactivity rather a decrease of reactivity was found. Copper has a smaller lattice constant than Pd and therefore lattice a compression effect lowers the energy of the center of the d-band.

ad 2) A model based on a spill-over effect can be considered, first described by Eikerling et al. [15]. Here, also the Au(111) support can directly be involved in the reaction pathway of the HER and does not act as inert support. Although the binding energy from hydrogen to Au(111) is weak compared to Pd or Pt, under certain circumstances a spill over of adsorbed hydrogen atoms to the Au(111) seems possible. The binding energy to Pd and Pt strongly decreases when the coverages of adsorbed hydrogen increases to almost a full monolayer which was supported by theoretical calculations [335]. Spill over can also be considered for HOR; calculations investigating this effect are under way.

ad 3) It was suggested from experiments on stepped Pt-surfaces by Löffler et al. [336] that a high electrocatalytic reactivity is caused by step sites. Results of Pd mono- and submonolayers on different vicinally stepped gold crystals investigated by Hernandez and Baltruschat [6, 7] underline the importance of steps and defect sites for HER activity. However, such dependence on the number of step sites was not directly found by Meier et al. [322] and Pandelov and Stimming [16]. Several new asprects highlight the extraordinary behavior of low coordinated atoms for the hydrogen reactions which were found in collaboration with theoretical groups during this thesis and partly described in **ad 1)**.

ad 4) An enhanced mass transport based on spherical diffusion in case of nanoparticles as compared to planar diffusion on extended surfaces may be considered as an explanation, too. Such enhanced mass transport was reported by Chen and Kucernak using microelectrodes [18] and analyzed by Quaino et al. [214]. Also Gasteiger et al. [4] and Neyerlin et al. [5] discussed enhanced diffusion in case of high particle dispersions and small particle sizes. Although this effect may play a role, it can not be the only reason since Pt sub-monolayers on HOPG yield considerably lower specific currents as compared to Au(111) as a substrate [20]. An explanation of the observed enhancement may be caused by any of the above factors or a combination of them; further experiments will clarify their respective importance.

The Volcano curve for hydrogen evolution correlates the hydrogen-metal bond strength with the exchange current density [212] and its maximum lies close to the value for HER on Pt electrodes which is approx. 1 mA/cm^2. Greeley et al. [337] and Nørskov et al. [335] complemented this Volcano curve using computational screening with metal overlayers and derived current densities which are about one order of magnitude higher as compared to the pure metal. Other reports have stated that the exchange current density j_0 for HER and HOR exceeds 25 mA/cm^2 for Pt/C [4] or is even in the range of 200-600 mA/cm^2 [5]. Previously reported values of j_0= 0.45-0.98 mA/cm^2 [226] or 1.7-3.0 mA/cm^2 [101] thus seem too low.

High exchange current densities of up to several 100 mA/cm^2 at Pt nanostructures on Au(111) and of up to several 10A/cm^2 at single Pt particles on an Au(111) surface for hydrogen reactions show a value in specific electrocatalytic activity of platinum that is more than two orders of magnitude higher as compared to a Pt(111) single crystal. The results show that the exchange current follows the same trend as reported in [4, 5] with values larger than 1 A/cm^2. Comparing this with the volcano, e.g. in ref. [337], indicates that either the volcano approach may need some modification for nanostructured surfaces or other, possibly experimental, effects yielded too low current densities in the past.

5.3.2 Geometric Current Density of HOR/HER of Pd and Pt on Au(111)

After a detailed discussion of the specific current density of Pd/Au(111) and Pt/Au(111) surfaces for hydrogen reactions, the geometric current density which represents the current through the whole electrode will be discussed. Although the determination of the geometric current density is done before calculating the specific current density the discussion afterwards will provide a closer look into the behavior of Pd and Pt catalysts. Additional results were taken from [16] and added to provide a complete overview in Figures 5.3 and 5.4.

In Figure 5.3 the hydrogen evolution for various overpotentials is plotted vs. the coverage of Pd or Pt on Au(111). The currents for the Pd/Au(111) (open symbols) systems were taken from [16] whereas the currents for the Pt/Au(111) (filled symbols) were obtained in this thesis. The overpotentials are between 50mV and 150mV to be sure that the reverse reaction does not interfere with the measured one. It is clearly seen that Pt/Au(111) electrodes are much more active as compared to Pd/Au(111). The difference is larger than one order of magnitude for high coverages and up to a factor of three for low coverages. It is also obvious that the geometric current density increases with decreasing Pd coverage whereas the geometric current density decreases with decreasing Pt coverage. This result shows a clear difference between Pd and Pt nanostructured Au(111) surfaces. Although a strain effect can be applied to both systems it can not be the only explanation due to the different behavior. Maybe diffusion effects play an important role since the activity of Pd is in each case lower and diffusion limitation of neither educts nor products is possibly not as crucial as it maybe in the platinum case. This means, the activity of Pt on Au(111) is in fact still higher, but measured currents can be limited by diffusion effects. Nevertheless, it is remarkable that the geometric current density for partly covered Pd covered Au(111) increases with a decrease of the active Pd surface area.

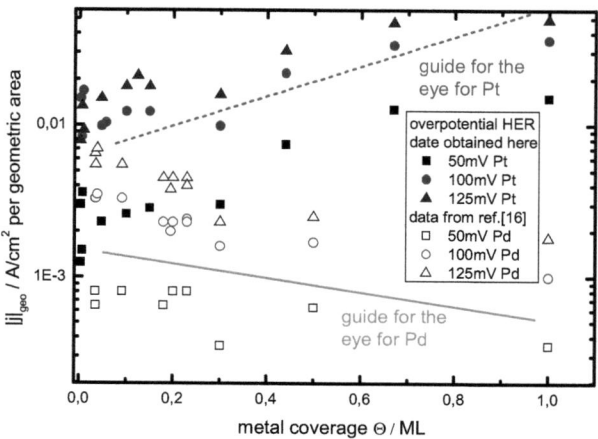

Figure 5. 3 Comparison of geometric current density for Pt/Au(111) and Pd/Au(111) for hydrogen evolution reaction for different overpotentials. Data for Pd/Au(111) are taken from [16].

A similar behavior is found for hydrogen oxidation reaction on Pd/Au(111) and Pt/Au(111) surfaces. The geometric current density was determined by potentiostatic pulses as described in Chapter 3. Although there are less data points, the trend is clearly visible in Figure 5.4 and shows as already seen for the HER a difference for Pd and Pt nanostructured Au(111) surfaces. Whereas Pt/Au(111) shows a decreasing geometrical current density for decreasing coverage, the Pd/Au(111) behave in the opposite way. Starting with a difference in absolute current densities of more than two orders of magnitude this difference is decreased to less than one order of magnitude for low coverages. This also leads to the conclusion that for low coverages the geometric current density for the different surfaces converges, as mentioned above it is not sufficient to only apply substrate effects. Diffusion effects are also one aspect which should be taken into account. But Pt on HOPG does not show increased reactivity although the very low Pt coverages were achieved and directly compared to the Pt/Au(111) system [20] indicating a minor influence due to diffusion in the investigated systems.

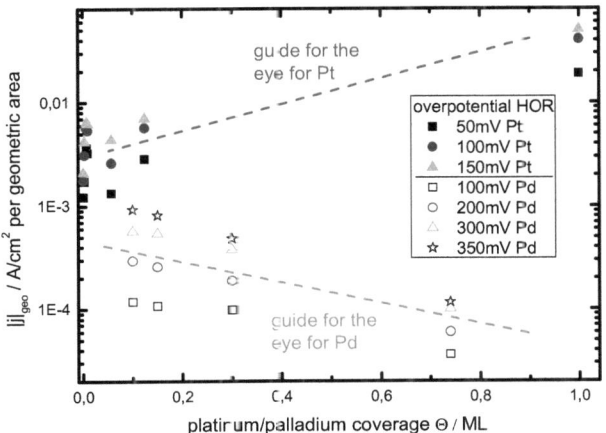

Figure 5.4 Comparison of geometric current density for Pt/Au(111) and Pd/Au(111) for hydrogen oxidation reaction at different overpotentials.

Up to now it is under discussion which model is able to explain the different behavior of the increasing geometric current density for Pd/Au(111) and decreasing geometric current density for Pt/Au(111) for hydrogen reactions. First results from cooperation with theoretical groups maintain the idea that the change of the binding energy between atomic hydrogen and Pd when deposited on Au(111) is shifted by more than 0.15eV compared to bulk Pd. According to the Vulcano plot in Figure 2.9 where Pd is locoated on the left side, Pd is shifted more to the top in the direction to Pt. Given that Pt is near the top of the Vulcano and thus offers almost perfect bindung conditions, the Au(111) support does affect the properties of Pt less.

5.3.3 Single Particles vs. Large Nanostructured Electrodes –Pt/Au(111)

In Chapter 4 large Pt nanostructured Au(111) electrodes and single Pt particles on Au(111) were shown in Figure 4.9 and Figure 4.12. It is assumed that the tip of the EC-STM above one insulated Pt particle on Au(111) probes a surface area of one μm^2. Then, coverage can be calculated and compared to the large nanostructured systems in the same plot. This was done in Figure 5.4 where the local approach to determine the activity of Pt is plotted together with the results obtained on large decorated electrodes. It can be seen that the insolated particles have an activity which is about four orders of magnitude higher as compared to bulk Pt(111) surfaces and at least one order of magnitude higher as compared to Pt nanostructured Au(111) although the coverage is less than one per mill (see Figure 5.5).

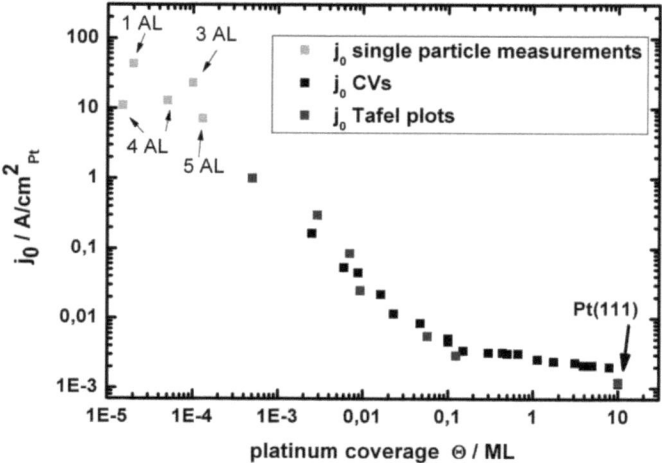

Figure 5. 5 Comparison of single particle measurements with large Pt nanostructured Au(111) electrodes.

It is also important to mention the enhancement of specific current density when lowering the coverage to less than 10% of a monolayer of Pt. With the assumption of a free area of one μm^2 Au(111) for one single Pt nanoparticle and the particle area a coverage in order to directly compare the results with large Pt nanostructured Au(111) surfaces was calculated. The results for single particles are plotted together with them of large structred Pt/Au(111) surfaces and continue the linear trend when lowering the amount of Pt.

This indicates a strong enhancement over a coverage variation of four orders of magnitudes. Due to the different morphologies of the particles the height is not proportional to the surface area of the particles. But, it is very important to mention that particles with one or three atomic layer height are more active compared to particles with four or five atomic layer height, which directly shows a strong influence of the support with a decreasing strain effect for increasing atomic layer heights (see Figure 5.6). Here, also the comparison to Pd nanoparticles on Au(111) is shown. The reactions rates for Pt nanoparticles were calculated with the exchange current densities from Figure 5.4 and the density of Pt atoms of $10^{15}/cm^2$. Values for the reaction rate for Pd nanoparticles were taken from [21]. Although the absolute values for the reaction rate are different the increase in activity with deacresing particle height is obvious. A similar slope indicates that the effect beyond the increase in activity is caused by the same reason which can also be transferred to behavior obtained on Pd and Pt mulitlayers which show also an increasing activity with decreasing thickness. The influence of diffusion in case of single particle measurements does not play a major role due to hemispherical diffusion to the nanoparticles. Overlapping diffusion effects which may arise from surrounding active catalyst particles are also absent.

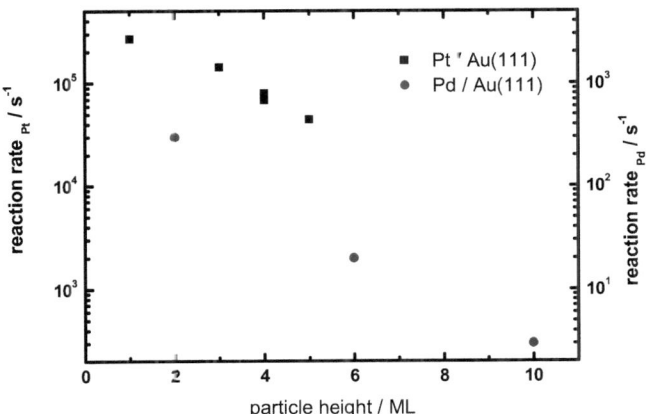

Figure 5. 6 Reaction rates for Pt and Pd single nanoparticles on Au(111).

The results show that both techniques, single particle measurements as well as large nanostructured electrodes based on the Pt/Au(111) and Pd/Au(111) system, show the same behavior. The important effects which describe the electrocatalytic behavior Pd and Pt overlayers on Au(111) were already discussed in Chapter 5 and can be obtained on single particles as well as large nanostrucutred surfaces.

5.3.4 Oxygen Reduction Reaction

Compared to the hydrogen reactions, the ORR does not show comparable effects on Pd/Au(111) and Pt/Au(111) electrodes; the specific current for this reaction is increasing for increasing Pd and Pt coverage on Au(111) in the monolayer regime and slightly increasing for decreasing amount of Pd and Pt in the submonolayer regime. An explanation of this effect can be caused by several factors including an influence of the support. Shao et al. [237] and Zhang et al. [338] showed an influence of the support when Pd and Pt monolayers were deposited on different metals. A strong decrease was found when depositing Pd and Pt on a support such as Ir(111), Au(111) and Rh(111) whereas a too strong or a too weak interaction to oxygen exists. This result was theoretically supported by Greeley et al. [43] showing that the position of the d-band center towards the Fermi level determines the interaction with the adsorbates. According to Sabatiers principle a neither too strong nor too weak binding is necessary to have a good catalyst. Here, it was shown that the electrocatalytic activity decreases in the following order: Pt(111) > multilayers of Pt on Au(111) > monolayer/submonolayers of Pt on Au(111). The effect of the support increases with decreasing layer thickness leading to changed binding energy to the adsorbates and thus decreased reactivity. Reducing the amount of Pt to less than a monolayer shows a small decrease in activity. Applying the calculations of Greeley and Nørskov [339] towards the oxygen binding energy to Pt for 1/3, 2/3 and a full monolayer of Pt on Au result in a consistent manner. Theoretical results show that with decreasing coverage the binding energy to oxygen is shifted to values which differ from the optimum binding energy and thus unbalance the Sabatier principle. This effect is also shown for the submonlayer regime below one full monolayer [339]. For high overpotentials at low coverages the results differ slightly from all other curves shown in Figure 4.23. Here an influence of different mass transport due to coverages smaller than 30% has to be considered which was also found from Watanabe et al. [340]. A detailed verification of this observed effect with mass transport correcting methods such as rotating disc electrode and flow cell can help to clarify the results in this boundary regime. Particle size effects for the ORR which may play role for small coverages do not play an important role which was recently published from Yano et al. [341].

5.3.5 Methanol Oxidation Reaction

The methanol oxidation was investigated using cyclic voltammograms and evaluated towards their maximum specific current density and their oxidation potential. Both results shown above in Figure 4.24 indicate that the Pt(111) reference shows the highest activity. The activity of the Pt mono- and submonolayers decreases with decreasing coverage. The maximum current density can be achieved if the Pt monolayer covers the whole surface. In the presented studies the current becomes constant for more than three monolayers. This indicates that a minimum thickness is necessary to obtain the current density compared to Pt(111). This was also found by Kim et al. [342] in studies of spontaneously deposited Pt islands on Au(111) whereas small and thin Pt deposits are disadvantageous for methanol oxidation. Also Tang et al. [343] found a similar result when investigating Au-Pt clusters. Increasing the content of Pt leads to an increasing activity towards the MOR. A shift of the peak potentials was also discussed and a value of $\Delta U = 200mV$ was found comparing bare Au and fully covered Pt samples. These data are in line with the results given in Chapter 4 which show $\Delta U = 150mV$ comparing 0.15ML of Pt on Au(111) and Pt(111). According to the discussion of HER/HOR and ORR the binding energy to methanol and the different reaction intermediates is an important factor for catalytic activity. Recent theoretical work from Ferrin and

Mavrikakis [261] shows the dependency of the orientation on the onset potential for the MOR on different metal supports. According to their work the (100) facet has a lower onset potential for the methanol oxidation for the direct mechanism compared to the (111) facet for the most metals such as Pt, Cu, Ag and Au. This result is in line with the data shown in Figure 4.24, whereas the roughened Pt(111) single crystal has a higher peak current density and a lower peak potential compared to the well prepared Pt(111) where a lower density of (100) sites is assumed. For very small coverages the peak potential is hardly measurable and a peak potential can not be accurately determined. Thus, Pt/Au(111) surfaces with coverages smaller than 0.15ML are not taken into account for investigating methanol oxidation. Investigations on Pd decorated Au(111) surfaces did not show any activity towards methanol oxidation, even at Pd multilayers no activity was measurable in contrast to the Pt/Au(111) system.

5.3.6 Summary of Support Effects

- The specific activity for HER/HOR increases with decreasing amount of noble metal for Pd/Au(111) and Pt/Au(111) nanostructred electrodes.
- Single Pt and Pd particles deposited on Au(111) increase their activity when the height of the particles is decreased.
- The geometric current density for HER/HOR increases with decreasing amount for Pd/Au(111) surfaces
- The geometric current density for the Pt/Au(111) increases with increasing amount of noble metal.
- The specific activity for ORR increases with increasing amount of noble metal for Pd/Au(111) and Pt/Au(111) nanostructured electrodes.
- The specific activity for MOR increases with increasing amount of noble metal for Pt/Au(111) nanostructured electrodes. The peak potential of the MOR is shifted to more negative potentials for higher coverages. There was no measurable activity found for Pd/Au(111) electrodes for MOR.
- The results for ORR and MOR are in contrast to the results obtained for the HER/HOR whereas the specific activity for hydrogen reactions increases with decreasing amount of noble metal on Pd/Au(111) and Pt/Au(111) nanostructured electrodes.

6 Summary and Conclusions

Substrate effects in electrocatalysis ranging from large nanostructured surfaces down to single particles deposited on single crystals for various reactions were the main topic of this thesis. An increased activity with decreasing amount of Pd and Pt deposited on Au(111) was found for hydrogen reactions; hydrogen evolution as well as hydrogen oxidation. In contrast to that result, the oxygen reduction and the methanol oxidation show an increasing activity with increasing amount of Pd and Pt. These findings were discussed in terms of low coordinated atoms, support effects, mass transport effects and a direct involvement of the support. It became clear, that the Au support plays a major role in catalytic behavior by causing a strain of the Pd and Pt overlayers.

In the discussion it was shown that several models obtained from different theoretical groups agree well with the experimental findings in this thesis. Density Functional Theory provides a theoretical insight into binding energies and proposed reaction pathways. Studies on bare single crystals, alloys of different metals and the influence of the support on foreign overlayers theoretically show a strong impact on the behavior of the catalysts deposited on various supports as compared to bare catalyst surfaces. Also a direct involvement of the support material as storage of intermediate reactions products seems possible. Reviewed systems from literature ranging from model surfaces to applied systems from experimental groups also show similar trends for the investigated hydrogen, oxygen and methanol reactions. Detailed studies on further model systems changing the support material such as the investigated ones will thus lead to a further insight into substrate effects and modifying catalytic activity specified on individual reactions.

A new kind of single crystalline surfaces were investigated with STM, SECPM, AFM and cyclic voltammetry in order to evaluate the quality and practicability of these surfaces for electrochemistry. All presented heteroepitaxially grown Ru, Rh and Ir single crystalline surfaces show single crystal quality. Therefore, they are suitable for a wide range of investigations under electrochemical conditions. Especially, the use as support material in electrocatalysis where foreign metals are deposited onto the substrates will clarify the influence of the support in more detail. Basic studies of Pt deposition on Ru(0001) and Cu-upd on Rh(111) in different electrolytes in this thesis have already underlined the applicability of these supports. During these studies the SECPM technique based on the potential difference between two electrodes in electrolytes was used. A resolution which is comparable to AFM and STM in air and under electrochemical conditions on the investigated surfaces was elaborated although the physical background is quite different. Potential distance spectroscopy with SECPM will also offer an insight into basic understanding of the electrochemical double layer which is of fundamental interest for the solid liquid interface.

A new local pH sensor in order to measure changes in proton concentration in the nanometerscale was developed. The successful use of a STM tip as a current sensor to locally measure single particle reactivity was the initial point to explore also a local potential technique. Based on the possibility to measure the OCP of the tip in the SECPM setup a potential sensitive tip was developed. Etched Pd tips with insulation to the very end of the apex and galvanostatically loaded with hydrogen acted as nanoelectrodes. These minimized electrodes are sensitive to proton concentration and thus change their OCP according to the pH of the solution surrounding the tip with nanometer resolution.

Summarized, the electrocatalytic activity on Pd and Pt nanostructured Au(111) surfaces towards hydrogen, oxygen and methanol reactions was studied. A strong influence of the support was found enhancing the hydrogen reactions and restraining the oxygen and methanol reactions. Also single Pt particles on Au(111) were generated and investigated towards their activity for hydrogen evolution showing an increase in activity with decreasing particle height. New single crystal supports were found to be excellent supports for electrochemical investigations. Principle studies regarding resolution and practicability of the SECPM were performed and a local potentiometric sensor for measurements of pH changes with a nanometer tip was developed.

7 References

[1] F. Maillard, M. Eikerling, O. V. Cherstiouk, S. Schreier, E. Savinova, and U. Stimming, *Faraday Discuss.*, **125** (2004) 357.
[2] F. Maillard, E. R. Savinova, and U. Stimming, *J. Electroanal. Chem.*, **599** (2007) 221.
[3] K. A. Friedrich, F. Henglein, U. Stimming, and W. Unkauf, *Electrochim. Acta*, **45** (2000) 3283.
[4] H. A. Gasteiger, J. E. Panels, and S. G. Yan, *J. Power Sources*, **127** (2004) 162.
[5] K. C. Neyerlin, W. B. Gu, J. Jorne, and H. A. Gasteiger, *J. Electrochem. Soc.*, **154** (2007) B631.
[6] F. Hernandez and H. Baltruschat, *Langmuir*, **22** (2006) 4877.
[7] F. Hernandez and H. Baltruschat, *J. Solid. State Chem.*, **11** (2007) 877.
[8] J. Greeley, T. F. Jaramillo, J. Bonde, I. B. Chorkendorff, and J. K. Nørskov, *Nat. Mater.*, **5** (2006) 909.
[9] J. R. Kitchin, J. K. Nørskov, M. A. Barteau, and J. G. Chen, *Phys. Rev. Lett.*, **93** (2004) 156801.
[10] J. K. Nørskov, T. Bligaard, A. Logadottir, J. R. Kitchin, J. G. Chen, S. Pandelov, and U. Stimming, *J. Electrochem. Soc.*, **152** (2005) J23.
[11] A. Roudgar and A. Gross, *Phys. Rev. B*, **67** (2003) 033409.
[12] A. Roudgar and A. Gross, *J. Electroanal. Chem.*, **548** (2003) 121.
[13] L. A. Kibler, A. M. El-Aziz, R. Hoyer, and D. M. Kolb, *Angew. Chem., Int. Ed.*, **44** (2005) 2080.
[14] J. Greeley, J. K. Nørskov, L. A. Kibler, A. M. El-Aziz, and D. M. Kolb, *ChemPhysChem*, **7** (2006) 1032.
[15] M. Eikerling, J. Meier, and U. Stimming, *Z. Phys. Chem. (Int. Ed.)*, **217** (2003) 395.
[16] S. Pandelov and U. Stimming, *Electrochim. Acta*, **52** (2007) 5548.
[17] L. A. Kibler, *Chemphyschem*, **7** (2006) 985.
[18] S. L. Chen and A. Kucernak, *J. Phys. Chem. B*, **108** (2004) 13984.
[19] J. Meier, K. A. Friedrich, and U. Stimming, *Faraday Discuss.*, **121** (2002) 365.
[20] T. Brülle and U. Stimming, *J. Electroanal. Chem.*, **636** (2009) 10.
[21] J. Meier, J. Schiotz, P. Liu, J. K. Nørskov, and U. Stimming, *Chem. Phys. Lett.*, **390** (2004) 440.
[22] T. Frelink, W. Visscher, and J. A. R. Vanveen, *J. Electroanal. Chem.*, **382** (1995) 65.
[23] K. Yahikozawa, Y. Fujii, Y. Matsuda, K. Nishimura, and Y. Takasu, *Electrochim. Acta*, **36** (1991) 973.
[24] K. Kinoshita, *J. Electrochem. Soc.*, **137** (1990) 845.
[25] Y. Takasu, N. Ohashi, X. G. Zhang, Y. Murakami, H. Minagawa, S. Sato, and K. Yahikozawa, *Electrochim. Acta*, **41** (1996) 2595.
[26] K. A. Friedrich, F. Henglein, U. Stimming, and W. Unkauf, *Colloids and Surf., A*, **134** (1998) 193.
[27] M. T. M. Koper, T. E. Shubina, and R. A. van Santen, *J. Phys. Chem. B*, **106** (2002) 686.
[28] Y. Y. Tong, C. Rice, A. Wieckowski, and E. Oldfield, *J. Am. Chem. Soc.*, **122** (2000) 1123.
[29] C. Rice, Y. Tong, E. Oldfield, A. Wieckowski, F. Hahn, F. Gloaguen, J. M. Leger, and C. Lamy, *J. Phys. Chem. B*, **104** (2000) 5803.
[30] B. Hammer and J. K. Nørskov, *Surf. Sci.*, **343** (1995) 211.
[31] J. K. Nørskov, *React. Kinet. Dev. Catal. Processes*, **122** (1999) 3.
[32] A. V. Ruban, H. L. Skriver, and J. K. Nørskov, *Phys. Rev. B*, **59** (1999) 15990.
[33] M. Mavrikakis, B. Hammer, and J. K. Nørskov, *Phys. Rev. Lett.*, **81** (1998) 2819.
[34] A. Ruban, B. Hammer, P. Stoltze, H. L. Skriver, and J. K. Nørskov, *J. Mol. Catal. A: Chem.*, **115** (1997) 421.
[35] B. Hammer and J. K. Nørskov, *Adv. Catal.*, **45** (2000) 71.
[36] J. Greeley, J. K. Nørskov, and M. Mavrikakis, *Annu. Rev. Phys. Chem.*, **53** (2002) 319.
[37] A. Roudgar and A. Gross, *Surf. Sci.*, **597** (2005) 42.
[38] E. Santos, P. Quaino, G. Soldano, and W. Schmickler, *J. Chem. Sci.*, **121** (2009) 575.
[39] E. Santos, A. Lundin, K. Potting, P. Quaino, and W. Schmickler, *J. Solid State Electrochem.*, **13** (2009) 1101.
[40] E. Santos, A. Lundin, K. Potting, P. Quaino, and W. Schmickler, *Phys. Rev. B*, **79** (2009)
[41] E. Santos, K. Potting, and W. Schmickler, *Faraday Discuss.*, **140** (2008) 209.
[42] E. Santos and W. Schmickler, *Chem. Phys.*, **332** (2007) 39.
[43] J. Greeley, J. K. Nørskov, and M. Mavrikakis, *Annual Review of Physical Chemistry*, **53** (2002) 319.
[44] M. Baldauf and D. M. Kolb, *Electrochim. Acta*, **38** (1993) 2145.
[45] M. Baldauf and D. M. Kolb, *J. Phys. Chem.*, **100** (1996) 11375.

[46] L. A. Kibler, A. M. El-Aziz, and D. M. Kolb, *J. Mol. Catal. A: Chem.*, **199** (2003) 57.
[47] H. Naohara, S. Ye, and K. Uosaki, *J. Electroanal. Chem.*, **500** (2001) 435.
[48] H. Naohara, S. Ye, and K. Uosaki, *Electrochim. Acta*, **45** (2000) 3305.
[49] G. Binnig, H. Rohrer, C. Gerber, and E. Weibel, *Phys. Rev. Lett.*, **49** (1982) 57.
[50] G. Binnig, H. Rohrer, C. Gerber, and E. Weibel, *Appl. Phys. Lett.*, **40** (1982) 178.
[51] W. Haiss, D. Lackey, J. K. Sass, and K. H. Besocke, *J. Chem. Phys.*, **95** (1991) 2193.
[52] P. K. Hansma and J. Tersoff, *J. Appl. Phys.*, **61** (1987) R1.
[53] R. Sonnenfeld and P. K. Hansma, *Science*, **232** (1986) 211.
[54] H.-J. Güntherodt and R. Wiesendanger, Scanning Tunneling Microscopy I, Springer Verlag, Berlin, 1994.
[55] H.-J. Güntherodt and R. Wiesendanger, Scanning Tunneling Microscopy II, Springer Verlag, Berlin, 1995.
[56] E. Meyer, H. J. Hug, and R. Bennewitz, Scanning Probe Microscopy, Springer Verlag, Berlin, 2003.
[57] C. Hamann and M. Hietschold, Raster-Tunnel-Mikroskopie, Akademie Verlag, Berlin, 1991.
[58] W. Schmickler, Interfacial Electrochemistry, Oxford University Press, Inc., New York, 1996.
[59] H. L. F. Helmholtz, *Annalen der Physik*, **89** (1853) 211.
[60] G. Gouy, *Journal de Physique et Le Radium*, **9** (1910) 457.
[61] D. L. Chapman, *Philosophical Magazine*, **25** (1913) 475.
[62] A. J. Bard and W. R. Faulkner, Electrochemical Methods: Fundamentals and Applications, Wiley, Weinheim, 2001.
[63] C. H. Hamann and W. Vielstich, Elektrochemie, WILEY-VCH Weinheim, 2005.
[64] Z. Borkowska and U. Stimming, in Structure of Electrified Interfaces, Frontiers of Electrochemistry, Vol. 2 (P. Lipkowski and P. N. Ross, eds.), VCH, 1993.
[65] R. Greef, R. Peat, L. M. Peter, D. Pletcher, and J. Robinson, Instrumental Methods in Electrochemistry, Southampton Electrochemsitry Group/Ellis Horwood Limited, Chichester, 1985.
[66] P. K. Hansma, V. B. Elings, O. Marti, and C. E. Bracker, *Science*, **242** (1988) 209.
[67] J. Frommer, *Angew. Chem., Int. Ed.*, **31** (1992) 1298.
[68] P. Samori, *J. Mater. Chem.*, **14** (2004) 1353.
[69] K. Itaya, *Prog. Surf. Sci.*, **58** (1998) 121.
[70] D. M. Kolb and F. C. Simeone, *Electrochim. Acta*, **50** (2005) 2989.
[71] H. J. Güntherodt and R. Wiesendanger, Scanning Tunneling Microscopy Volume I - III, Springer, Berlin, 1993.
[72] A. J. Bard and F. R. Fan, *Faraday Discuss.*, **94** (1992) 1.
[73] P. Lustenberger, H. Rohrer, R. Christoph, and H. Siegenthaler, *J. Electroanal. Chem.*, **243** (1988) 225.
[74] K. Itaya and E. Tomita, *Surf. Sci.*, **201** (1988) L507.
[75] G. Binnig, N. Garcia, H. Rohrer, J. M. Soler, and F. Flores, *Phys. Rev. B*, **30** (1984) 4816.
[76] G. E. Engelmann, J. C. Ziegler, and D. M. Kolb, *Surf. Sci.*, **401** (1998) L420.
[77] S. Pandelov, PhD Thesis, Technische Universität München, (München), 2007.
[78] G. Binnig, C. F. Quate, and C. Gerber, *Phys. Rev. Lett.*, **56** (1986) 930.
[79] G. Meyer and N. M. Amer, *Appl. Phys. Lett.*, **53** (1988) 1045.
[80] F. J. Giessibl, *Rev. Mod. Phys.*, **75** (2003) 949.
[81] R. Garcia and R. Perez, *Surf. Sci. Rep.*, **47** (2002) 197.
[82] W. A. Hofer, A. S. Foster, and A. L. Shluger, *Rev. Mod. Phys.*, **75** (2003) 1287.
[83] D. H. Woo, J. S. Yoo, S. M. Park, I. C. Jeon, and H. Kang, *Bull. Korean Chem. Soc.*, **25** (2004) 577.
[84] C. Hurth, C. Z. Li, and A. J. Bard, *J. Phys. Chem. C*, **111** (2007) 4620.
[85] Y. T. Kim and A. J. Bard, *Langmuir*, **8** (1992) 1096.
[86] H. P. Chang and A. J. Bard, *J. Am. Chem. Soc.*, **113** (1991) 5588.
[87] D. W. Suggs and A. J. Bard, *J. Phys. Chem.*, **99** (1995) 8349.
[88] C. Corbella, E. Pascual, G. Oncins, C. Canal, J. L. Andujar, and E. Bertran, *Thin Solid Films*, **482** (2005) 293.
[89] C. Baier and U. Stimming, *Angew. Chem., Int. Ed.*, **48** (2009) 5342.
[90] H. Ohshima, *Colloid Polym. Sci.*, **252** (1974) 158.
[91] J. E. Sader and A. M. Lenhoff, *J. Colloid Interface Sci.*, **201** (1998) 233.
[92] J. E. Sader and D. Y. C. Chan, *J. Colloid Interface Sci.*, **218** (1999) 423.
[93] R. F. Hamou, P. U. Biedermann, M. Rohwerder, and A. T. Blumenau, *Excerpt from the Proceedings of the COMSOL Conference 2008 Hannover*, (2009)

[94] R. F. Hamou, P. U. Biedermann, A. Erbe, and M. Rohwerder, *Electrochim. Acta,* **55** (2010) 5210.
[95] C. Baier, PhD Thesis, Technische Universität München, (München), 2010.
[96] N. M. Markovic, S. T. Sarraf, H. A. Gasteiger, and P. N. Ross, *J. Chem. Soc. - Faraday Trans.,* **92** (1996) 3719.
[97] N. M. Markovic, B. N. Grgur, and P. N. Ross, *J. Phys. Chem. B,* **101** (1997) 5405.
[98] J. Barber, S. Morin, and B. E. Conway, *J. Electroanal. Chem.,* **446** (1998) 125.
[99] J. Perez, E. R. Gonzalez, and H. M. Villullas, *J. Phys. Chem. B,* **102** (1998) 10931.
[100] J. H. Barber and B. E. Conway, *J. Electroanal. Chem.,* **461** (1999) 80.
[101] K. Seto, A. Iannelli, B. Love, and J. Lipkowski, *J. Electroanal. Chem.,* **226** (1987) 351.
[102] S. Schuldin, M. Rosen, and D. R. Flinn, *J. Electrochem. Soc.,* **117** (1970) 1251.
[103] H. Kita, S. Ye, and Y. Gao, *J. Electroanal. Chem.,* **334** (1992) 351.
[104] R. Gomez, A. Fernandezvega, J. M. Feliu, and A. Aldaz, *J. Phys. Chem.,* **97** (1993) 4769.
[105] L. A. Kibler, Preparation and Characterization of Noble Metal Single Crystal Electrode Surfaces, International Society of Electrochemistry, 2003.
[106] J. Clavilier, R. Faure, G. Guinet, and R. Durand, *J. Electroanal. Chem.,* **107** (1980) 205.
[107] M. P. Soriaga, *Prog. Surf. Sci.,* **39** (1992) 325.
[108] A. T. Hubbard, *Chem. Rev.,* **88** (1988) 633.
[109] T. L. Chen, X. M. Li, S. Zhang, and X. Zhang, *Appl. Phys. A,* **80** (2005) 73.
[110] D. Akai, K. Hirabayashi, M. Yokawa, K. Sawada, and M. Ishida, *J. Cryst. Growth,* **264** (2004) 463.
[111] S. Gsell, M. Fischer, R. Brescia, M. Schreck, P. Huber, F. Bayer, B. Stritzker, and D. G. Schlom, *Appl. Phys. Lett.,* **91** (2007) 061501.
[112] S. Gsell, T. Bauer, J. Goldfuss, M. Schreck, and B. Stritzker, *Appl. Phys. Lett.,* **84** (2004) 4541.
[113] J. Clavilier, *J. Electroanal. Chem.,* **107** (1980) 211.
[114] S. Motoo and N. Furuya, *J. Electroanal. Chem.,* **172** (1984) 339.
[115] N. Markovic, M. Hanson, G. McDougall, and E. Yeager, *J. Electroanal. Chem.,* **214** (1986) 555.
[116] S. Motoo and N. Furuya, *Berichte Der Bunsen-Gesellschaft,* **91** (1987) 457.
[117] J. Clavilier, A. Rodes, K. Elachi, and M. A. Zamakhchari, *Journal De Chimie Physique Et De Physico-Chimie Biologique,* **88** (1991) 1291.
[118] J. Clavilier, J. M. Orts, and J. M. Feliu, *Journal De Physique Iv,* **4** (1994) 303.
[119] J. M. Feliu, A. Rodes, J. M. Orts, and J. Clavilier, *Pol. J. Chem.,* **68** (1994) 1575.
[120] L. A. Kibler, A. Cuesta, M. Kleinert, and D. M. Kolb, *J. Electroanal. Chem.,* **484** (2000) 73.
[121] J. Clavilier, J. M. Feliu, A. Fernandezvega, and A. Aldaz, *J. Electroanal. Chem.,* **269** (1989) 175.
[122] J. Clavilier, D. Armand, S. G. Sun, and M. Petit, *J. Electroanal. Chem.,* **205** (1986) 267.
[123] A. M. Funtikov, U. Linke, U. Stimming, and R. Vogel, *Surf. Sci.,* **324** (1995) L343.
[124] L. A. Kibler, A. Cuesta, M. Kleinert, and D. M. Kolb, *Journal of Electroanalytical Chemistry,* **484** (2000) 73.
[125] S. Horch, H. T. Lorensen, S. Helveg, E. Laegsgaard, I. Stensgaard, K. W. Jacobsen, J. K. Nørskov, and F. Besenbacher, *Nature,* **398** (1999) 134.
[126] N. M. Markovic, B. N. Grgur, C. A. Lucas, and P. N. Ross, *Surf. Sci.,* **384** (1997) L805.
[127] E. Vlieg, I. K. Robinson, and K. Kern, *Surf. Sci.,* **233** (1990) 248.
[128] D. A. Scherson and D. M. Kolb, *J. Electroanal. Chem.,* **176** (1984) 353.
[129] A. M. Funtikov, U. Stimming, and R. Vogel, *J. Electroanal. Chem.,* **428** (1997) 147.
[130] B. Braunschweig and W. Daum, *Langmuir,* **25** (2009) 11112.
[131] U. Frese, T. Iwasita, W. Schmickler, and U. Stimming, *J. Phys. Chem.,* **89** (1985) 1059.
[132] U. Frese and U. Stimming, *J. Electroanal. Chem.,* **198** (1986) 409.
[133] N. M. Markovic and P. N. Ross, *Surf. Sci. Rep.,* **45** (2002) 121.
[134] N. M. Markovic, B. N. Grgur, and P. N. Ross, *Journal of Physical Chemistry B,* **101** (1997) 5405.
[135] J. Maruyama, M. Inaba, K. Katakura, Z. Ogumi, and Z.-i. Takehara, *J. Electroanal. Chem.,* **447** (1998) 201.
[136] V. S. Bagotzky and N. V. Osetrova, *J. Electroanal. Chem.,* **43** (1973) 233.
[137] W. Vogel, L. Lundquist, P. Ross, and P. Stonehart, *Electrochim. Acta,* **20** (1975) 79.
[138] C. G. Zoski, *J. Phys. Chem. B,* **107** (2003) 6401.
[139] J. Zhou, Y. Zu, and A. J. Bard, *J. Electroanal. Chem.,* **491** (2000) 22.
[140] M. H. Dishner, M. M. Ivey, S. Gorer, J. C. Hemminger, and F. J. Feher, *J. Vac. Sci. Technol., A,* **16** (1998) 3295.
[141] D. M. Kolb, *Prog. Surf. Sci.,* **51** (1996) 109.
[142] O. M. Magnussen, J. Hageböck, J. Hotlos, and R. J. Behm, *Faraday Discuss.,* **94** (1992) 329.

[143] T. Dretschkow and T. Wandlowski, *Phys. Chem. Chem. Phys,* **101** (1997) 749.
[144] G. J. Edens, X. Gao, and M. J. Weaver, *J. Electroanal. Chem.,* **375** (1994) 357.
[145] X. P. Gao, A. Hamelin, and M. J. Weaver, *Phys. Rev. B,* **44** (1991) 10983.
[146] X. P. Gao, A. Hamelin, and M. J. Weaver, *Phys. Rev. Lett.,* **67** (1991) 618.
[147] O. M. Magnussen, J. Hotlos, R. J. Behm, N. Batina, and D. M. Kolb, *Surf. Sci.,* **296** (1993) 310.
[148] O. M. Magnussen, J. Wiechers, and R. J. Behm, *Surf. Sci.,* **289** (1993) 139.
[149] M. Sturmat, R. Koch, and K. H. Rieder, *Phys. Rev. Lett.,* **77** (1996) 5071.
[150] A. S. Dakkouri, *Solid State Ionics* **94** (1997) 99.
[151] A. Hamelin, *J. Electroanal. Chem.,* **407** (1996) 1.
[152] G. J. Brug, M. Sluytersrehbach, J. H. Sluyters, and A. Hamelin *J. Electroanal. Chem.,* **181** (1984) 245.
[153] A. Hamelin and M. J. Weaver, *J. Electroanal. Chem.,* **223** (1987) 171.
[154] A. Hamelin, S. Röttgermann, and W. Schmickler, *J. Electroanal. Chem.,* **230** (1987) 281.
[155] S. Trasatti, *Russ. J. Electrochem.,* **41** (2005) 1255.
[156] B. Alvarez, V. Climent, A. Rodes, and J. M. Feliu, *Phys. Chem. Chem. Phys.,* **3** (2001) 3269.
[157] R. Hoyer, L. A. Kibler, and D. M. Kolb, *Electrochim. Acta,* **49** (2003) 63.
[158] M. Arenz, V. Stamenkovic, P. N. Ross, and N. M. Markovic, *Surf. Sci.,* **573** (2004) 57.
[159] A. M. El-Aziz, R. Hoyer, L. A. Kibler, and D. M. Kolb, *Electrochim. Acta,* **51** (2006) 2518.
[160] Y. Pluntke, L. A. Kibler, and D. M. Kolb, *Phys. Chem. Chem. Phys.,* **10** (2008) 3684.
[161] L. A. Kibler, M. Kleinert, V. Lazarescu, and D. M. Kolb, *Surf. Sci.,* **498** (2002) 175.
[162] L. A. Kibler, M. Kleinert, and D. M. Kolb, *Surf. Sci.,* **461** (2000) 155.
[163] L. A. Kibler, M. Kleinert, R. Randler, and D. M. Kolb, *Surf. Sci.,* **443** (1999) 19.
[164] J. Clavilier, M. Wasberg, M. Petit, and L. H. Klein, *J. Electroanal. Chem.,* **374** (1994) 123.
[165] L. J. Wan, S. L. Yau, G. M. Swain, and K. Itaya, *J. Electroanal. Chem.,* **381** (1995) 105.
[166] Y. E. Sung, S. Thomas, and A. Wieckowski, *J. Phys. Chem.,* **99** (1995) 13513.
[167] R. Gomez, J. M. Orts, J. M. Feliu, J. Clavilier, and L. H. Klein, *J. Electroanal. Chem.,* **432** (1997) 1.
[168] P. Zelenay, G. Horanyi, C. K. Rhee, and A. Wieckowski, *J. Electroanal. Chem.,* **300** (1991) 499.
[169] L. J. Wan, S. L. Yau, and K. Itaya, *J. Phys. Chem.,* **99** (1995) 9507.
[170] C. K. Rhee, M. Wasberg, G. Horanyi, and A. Wieckowski, *J. Electroanal. Chem.,* **291** (1990) 281.
[171] S. L. Yau, Y. G. Kim, and K. Itaya, *J. Am. Chem. Soc.,* **118** (1996) 7795.
[172] S. Motoo and N. Furuya, *J. Electroanal. Chem.,* **167** (1984) 309.
[173] S. Motoo and N. Furuya, *J. Electroanal. Chem.,* **181** (1984) 301.
[174] M. A. Vanhove, R. J. Koestner, P. C. Stair, J. P. Biberian, L. L. Kesmodel, I. Bartos, and G. A. Somorjai, *Surf. Sci.,* **103** (1981) 139.
[175] M. A. Vanhove, R. J. Koestner, P. C. Stair, J. P. Biberian, L. L. Kesmodel, I. Bartos, and G. A. Somorjai, *Surf. Sci.,* **103** (1981) 218.
[176] C. M. Chan and M. A. Vanhove, *Surf. Sci.,* **171** (1986) 226.
[177] R. Gomez and M. J. Weaver, *Langmuir,* **18** (2002) 4426.
[178] T. Pajkossy, L. A. Kibler, and D. M. Kolb, *J. Electroanal. Chem,* **582** (2005) 69.
[179] T. Pajkossy, L. A. Kibler, and D. M. Kolb, *J. Electroanal. Chem,* **600** (2007) 113.
[180] W. B. Wang, M. S. Zei, and G. Ertl, *Chem. Phys. Lett.,* **355** (2002) 301.
[181] W. B. Wang, M. S. Zei, and G. Ertl, *Phys. Chem. Chem. Phys.,* **3** (2001) 3307.
[182] A. Bergbreiter, A. Berko, P. M. Erne, H. E. Hoster, and R. J. Behm, *Vacuum,* **84** (2009) 13.
[183] H. Hartmann, T. Diemant, J. Bansmann, and R. J. Behm, *Surf. Sci.,* **603** (2009) 1456.
[184] H. Rauscher, T. Hager, T. Diemant, H. Hoster, F. B. De Mongeot, and R. J. Behm, *Surf. Sci.,* **601** (2007) 4608.
[185] H. Hoster, B. Richter, and R. J. Behm, *J. Phys. Chem. B,* **108** (2004) 14780.
[186] A. M. El-Aziz and L. A. Kibler, *Electrochem. Commun.,* **4** (2002) 866.
[187] H. Inoue, J. X. Wang, K. Sasaki, and R. R. Adzic, *J. Electroanal. Chem.,* **554-555** (2003) 77.
[188] A. M. El-Aziz and L. A. Kibler, *Electrochem. Commun.,* **4** (2002) 866.
[189] N. S. Marinkovic, J. X. Wang, H. Zajonz, and R. R. Adzic, *J. Electroanal. Chem.,* **500** (2001) 388.
[190] J. V. Zoval, R. M. Stiger, P. R. Biernacki, and R. M. Penner, *J. Phys. Chem.,* **100** (1996) 837.
[191] J. V. Zoval, J. Lee, S. Gorer, and R. M. Penner, *J. Phys. Chem. B,* **102** (1998) 1166.
[192] F. Gloaguen, J. M. Léger, C. Lamy, A. Marmann, U. Stimming, and R. Vogel, *Electrochim. Acta,* **44** (1999) 1805.
[193] E. Budevski, G. Staikov, and W. J. Lorenz, Electrochemical Phase Formation and Growth An Introduction to the Initial Stage of Metal Depostion, WILEY-VCH, Weinheim, 1996.

[194] E. Budevski, G. Staikov, and W. J. Lorenz, *Electrochim. Acta,* **45** (2000) 2559.
[195] M. Petri and D. M. Kolb, *Phys. Chem. Chem. Phys.,* **4** (2002) 1211.
[196] Z. X. Xie and D. M. Kolb, *J. Electroanal. Chem.,* **481** (2000) 177.
[197] S. G. Garcia, D. R. Salinas, C. E. Mayer, W. J. Lorenz, and G. Staikov, *Electrochim. Acta,* **48** (2003) 1279.
[198] R. Schuster, V. Kirchner, P. Allongue, and G. Ertl, *Science,* **289** (2000) 98.
[199] V. Kirchner, L. Cagnon, R. Schuster, and G. Ertl, *Appl. Phys. Lett.,* **79** (2001) 1721.
[200] W. Schindler, D. Hofmann, and J. Kirschner, *J. Appl. Phys.,* **87** (2000) 7007.
[201] R. T. Potzschke, G. Staikov, W. J. Lorenz, and W. Wiesbeck, *J. Electrochem. Soc.,* **146** (1999) 141.
[202] W. Schindler, D. Hofmann, and J. Kirschner, *J. Electrochem. Soc.,* **148** (2001) C124.
[203] A. J. Bard and M. V. Mirkin, Scanning Electrochemical Microscopy, John Wiley & Sons, New York, 2001.
[204] N. Baltes, L. Thouin, C. Amatore, and J. Heinze, *Angew. Chem., Int. Ed.,* **43** (2004) 1431.
[205] M. G. Del Popolo, E. P. M. Leiva, H. Kleine, J. Meier, U. Stimming, M. Mariscal, and W. Schmickler, *Electrochim. Acta,* **48** (2003) 1287.
[206] D. M. Kolb, R. Ullmann, and T. Will, *Science,* **275** (1997) 1097.
[207] D. M. Kolb, R. Ullmann, and J. C. Ziegler, *Electrochim. Acta,* **43** (1998) 2751.
[208] M. Del Popolo, E. Leiva, H. Kleine, J. Meier, U. Stimming, M. Mariscal, and W. Schmickler, *Appl. Phys. Lett.,* **81** (2002) 2635.
[209] T. Erdey-Gruz and M. Volmer, *Z. Phys. Chem.,* **150** (1930) 203.
[210] J. Tafel, *Z. Phys. Chem.,* **50** (1905) 641.
[211] J. Heyrovsky, *Recueil Des Travaux Chimiques Des Pays-Bas,* **46** (1927) 582.
[212] S. Trasatti, *J. Electroanal. Chem.,* **39** (1972) 163.
[213] R. Parsons, *Trans. Faraday Soc.,* **54** (1958) 1053.
[214] P. M. Quaino, J. L. Fernandez, M. R. G. de Chialvo, and A. C. Chialvo, *J. Mol. Catal. A: Chem.,* **252** (2006) 156.
[215] A. Gross, *Top. Catal.,* **37** (2006) 29.
[216] B. E. Conway and J. O. Bockris, *J. Chem. Phys.,* **26** (1957) 532.
[217] L. I. Krishtalik and P. Delahay, eds., Advances in Electrochemistry and Electrochemical Engineering, Interscience, New York, 1970.
[218] M. R. Tarasevich, A. Sadkowski, and E. Yeager, in Comprehensive Treatise of Electrochemistry (J. O. Bockris, E. E. Conway, E. Yeager, S. U. M. Khan, and R. E. White, eds.), Plenum, New York, 1983.
[219] N. M. Markovic, T. J. Schmidt, V. Stamenkovic, and P. N. Ross, *Fuel Cells,* **1** (2001) 105.
[220] H. S. Wroblowa, Y. C. Pan, and G. Razumney, *J. Electroanal. Chem.,* **69** (1976) 195.
[221] V. S. Bagotskii, M. R. Tarasevich, and V. Y. Filinovskii, *Elektrokhimiya,* **5** (1969) 1218.
[222] E. Yeager, *Electrochim. Acta,* **29** (1984) 1527.
[223] E. Yeager, *J. Mol. Catal.,* **38** (1986) 5.
[224] N. M. Markovic, R. R. Adzic, B. D. Cahan, and E. B. Yeager, *J. Electroanal. Chem.,* **377** (1994) 249.
[225] N. M. Markovic, H. A. Gasteiger, and P. N. Ross, *J. Phys. Chem.,* **99** (1995) 3411.
[226] B. N. Grgur, N. M. Markovic, and P. N. Ross, *Can. J. Chem.,* **75** (1997) 1465.
[227] U. A. Paulus, A. Wokaun, G. G. Scherer, T. J. Schmidt, V. Stamenkovic, N. M. Markovic, and P. N. Ross, *Electrochim. Acta,* **47** (2002) 3787.
[228] M. H. Shao, P. Liu, J. L. Zhang, and R. Adzic, *J. Phys. Chem. B,* **111** (2007) 6772.
[229] V. Stamenkovic, B. S. Mun, K. J. J. Mayrhofer, P. N. Ross, N. M. Markovic, J. Rossmeisl, J. Greeley, and J. K. Nørskov, *Angew. Chem., Int. Ed.,* **45** (2006) 2897.
[230] T. Toda, H. Igarashi, H. Uchida, and M. Watanabe, *J. Electrochem. Soc.,* **146** (1999) 3750.
[231] T. J. Schmidt, V. Stamenkovic, M. Arenz, N. M. Markovic, and P. N. Ross, *Electrochimica Acta,* **47** (2002) 3765.
[232] K. Sasaki, Y. Mo, J. X. Wang, M. Balasubramanian, F. Uribe, J. McBreen, and R. R. Adzic, *Electrochim. Acta,* **48** (2003) 3841.
[233] V. Stamenkovic, B. S. Mun, K. J. J. Mayrhofer, P. N. Ross, N. M. Markovic, J. Rossmeisl, J. Greeley, and J. K. Nørskov, *Angew. Chem. Int. Edn,* **45** (2006) 2897.
[234] V. R. Stamenkovic, B. Fowler, B. S. Mun, G. F. Wang, P. N. Ross, C. A. Lucas, and N. M. Markovic, *Science,* **315** (2007) 493.

[235] V. R. Stamenkovic, B. S. Mun, M. Arenz, K. J. J. Mayrhofer, C. A. Lucas, G. F. Wang, P. N. Ross, and N. M. Markovic, *Nat. Mater.*, **6** (2007) 241.
[236] T. J. Schmidt, V. Stamenkovic, M. Arenz, N. M. Markovic, and P. N. Ross, *Electrochim. Acta*, **47** (2002) 3765.
[237] M. H. Shao, T. Huang, P. Liu, J. Zhang, K. Sasaki, M. B. Vukmirovic, and R. R. Adzic, *Langmuir*, **22** (2006) 10409.
[238] M. H. Shao, P. Liu, J. L. Zhang, and R. Adzic, *Journal of Physical Chemistry B*, **111** (2007) 6772.
[239] J. Greeley, I. E. L. Stephens, A. S. Bondarenko, T. P. Johansson, H. A. Hansen, T. F. Jaramillo, J. Rossmeisl, I. Chorkendorff, and J. K. Nørskov, *Nat. Chem.*, **1** (2009) 552.
[240] R. W. Reeve, P. A. Christensen, A. Hamnett, S. A. Haydock, and S. C. Roy, *J. Electrochem. Soc.*, **145** (1998) 3463.
[241] R. J. Allen, D. Czerwiec, J. R. Giallombardo, and K. Shaikh, Vol. 5 958 197 (US-Patent, ed.), 1998.
[242] R. J. Allen, J. R. Giallombardo, D. Czerwiec, E. S. De Castros, and K. Shaikh, Vol. 6 149 782 (US-Patent, ed.), 2000.
[243] B. Bittinscattaneo, S. Wasmus, B. Lopezmishima, and W. Vielstich, *J. Appl. Electrochem.*, **23** (1993) 625.
[244] R. Holze, I. Vogel, and W. Vielstich, *J. Electrochem. Soc.*, **133** (1986) C115.
[245] G. Faubert, R. Cote, J. P. Dodelet, M. Lefevre, and P. Bertrand, *Electrochim. Acta*, **44** (1999) 2589.
[246] G. Faubert, R. Cote, D. Guay, J. P. Dodelet, G. Denes, and P. Bertrand, *Electrochim. Acta*, **43** (1998) 341.
[247] P. Gouerec and M. Savy, *Electrochim. Acta*, **44** (1999) 2653.
[248] P. Gouerec, M. Savy, and J. Riga, *Electrochim. Acta*, **43** (1998) 743.
[249] U. A. Paulus, T. J. Schmidt, and H. Gasteiger, in Handbook of Fuel Cell Technology (W. Vielstich and H. A. Gasteiger, eds.), John Wiley & Sons, New York, 2003.
[250] Y.-C. Lu, H. A. Gasteiger, M. C. Parent, V. Chiloyan, and Y. Shao-Horn, *Electrochem. Solid State Lett.*, **13** (2010) A69.
[251] Y. C. Lu, Z. Xu, H. A. Gasteiger, S. Chen, K. Hamad-Schifferli, and Y. Shao-Horn, *J. Am. Chem. Soc.*, **132** (2010) 12170.
[252] M. J. Janik, C. D. Taylor, and M. Neurock, *Top. Catal.*, **46** (2007) 306.
[253] T. Iwasita, *Electrochim. Acta*, **47** (2002) 3663.
[254] R. Parsons and T. Vanderncot, *J. Electroanal. Chem.*, **257** (1988) 9.
[255] E. Herrero, K. Franaszczuk, and A. Wieckowski, *J. Phys. Chem.*, **98** (1994) 5074.
[256] S. Wasmus and A. Kuver, *J. Electroanal. Chem.*, **461** (1999) 14.
[257] H. A. Gasteiger, N. Markovic, P. N. Ross, and E. J. Cairns, *J. Phys. Chem.*, **97** (1993) 12020.
[258] A. Hamnett, *Catal. Today*, **33** (1997) 445.
[259] D. Cao, G. Q. Lu, A. Wieckowski, S. A. Wasileski, and M. Neurock, *J. Phys. Chem. B*, **109** (2005) 11622.
[260] J. D. Lovic, A. V. Tripkovic, S. L. J. Gojkovic, K. D. Popovic, D. V. Tripkovic, P. Olszewski, and A. Kowal, *J. Electroanal. Chem.*, **581** (2005) 294.
[261] P. Ferrin and M. Mavrikakis, *J. Am. Chem. Soc.*, **131** (2009) 14381.
[262] P. Ferrin, A. U. Nilekar, J. Greeley, M. Mavrikakis, and J. Rossmeisl, *Surf. Sci.*, **602** (2008) 3424.
[263] J. Greeley and M. Mavrikakis, *J. Am. Chem. Soc.*, **126** (2004) 3910.
[264] S. K. Desai, M. Neurock, and K. Kourtakis, *J. Phys. Chem. B*, **106** (2002) 2559.
[265] J. Greeley and M. Mavrikakis, *J. Am. Chem. Soc.*, **124** (2002) 7193.
[266] C. Hartnig, J. Grimminger, and E. Spohr, *Electrochim. Acta*, **52** (2007) 2236.
[267] G. T. Burstein, C. J. Barnett, A. R. Kucernak, and K. R. Williams, *Catal. Today*, **38** (1997) 425.
[268] E. A. Batista and T. Iwasita, *Langmuir*, **22** (2006) 7912.
[269] E. A. Batista, G. R. P. Malpass, A. J. Motheo, and T. Iwasita, *Electrochem. Commun.*, **5** (2003) 843.
[270] Y. X. Chen, M. Heinen, Z. Jusys, and R. J. Behm, *Langmuir*, **22** (2006) 10399.
[271] H. Hoster, T. Iwasita, H. Baumgartner, and W. Vielstich, *Phys. Chem. Chem. Phys.*, **3** (2001) 337.
[272] T. H. M. Housmans and M. T. M. Koper, *J. Phys. Chem. B*, **107** (2003) 8557.
[273] T. H. M. Housmans, A. H. Wonders, and M. T. M. Koper, *J. Phys. Chem. B*, **110** (2006) 10021.
[274] D. Kardash, C. Korzeniewski, and N. Markovic, *J. Electroanal. Chem.*, **500** (2001) 518.
[275] T. D. Jarvi and E. M. Stuve, in The Science of Electrocatalysis on Bimetallic Surfaces (J. Lipkowski and P. Ross, eds.), Wiley-VCH, New York, 1998, p. 75.
[276] E. Reddington, A. Sapienza, B. Gurau, R. Viswanathan, S. Sarangapani, E. S. Smotkin, and T. E. Mallouk, *Science*, **280** (1998) 1735.

[277] O. A. Petrii, *J. Solid State Electrochem.*, **12** (2008) 609.
[278] A. Lima, C. Coutanceau, J. M. Leger, and C. Lamy, *J. Appl. Electrochem.*, **31** (2001) 379.
[279] P. Strasser, Q. Fan, M. Devenney, W. H. Weinberg, P. Liu, and J. K. Nørskov, *J. Phys. Chem. B*, **107** (2003) 11013.
[280] K. W. Park, J. H. Choi, B. K. Kwon, S. A. Lee, Y. E. Sung, H. Y. Ha, S. A. Hong, H. Kim, and A. Wieckowski, *J. Phys. Chem. B*, **106** (2002) 1869.
[281] J. Greeley and M. Mavrikakis, *Catal. Today*, **111** (2006) 52.
[282] M. Watanabe and S. Motoo, *J. Electroanal. Chem.*, **60** (1975) 267.
[283] M. Watanabe and S. Motoo, *J. Electroanal. Chem.*, **60** (1975) 275.
[284] J. Meier, PhD Thesis, Technische Universität München, (München), 2003.
[285] P. Hugelmann, PhD Thesis, Universität Karlsruhe, (Karlsruhe), 2005.
[286] A. J. Bard, R. Parsons, and J. Jordan, Standard potentials in aqueous solutuion, Marcel Dekker, New York, 1985.
[287] G. J. Hills and D. J. G. Ives, Reference Electrodes, Theory and Practice, Academic Press, London, 1961.
[288] D. Midgley, *Talanta*, **37** (1990) 767.
[289] T. B. Flanagan and F. A. Lewis, *Trans. Faraday Soc.*, **55** (1959) 1400.
[290] D. D. Macdonald, P. R. Wentrcek, and A. C. Scott, *J. Electrochem. Soc.*, **127** (1980) 1745.
[291] M. J. Vasile and C. G. Enke, *J. Electrochem. Soc.*, **112** (1965) 865.
[292] R. C. Wolfe, K. G. Weil, B. A. Shaw, and H. W. Pickering, *J. Electrochem. Soc.*, **152** (2005) B82.
[293] S. Schuldiner, G. Castellan, and J. P. Hoare, *J. Chem. Phys.*, **28** (1958) 16.
[294] D. Jung, Diploma Thesis, Technische Universität München, (München), 2004.
[295] I. Horcas, R. Fernandez, J. M. Gomez-Rodriguez, J. Colchero, J. Gomez-Herrero, and A. M. Baro, *Rev. Sci. Instrum.*, **78** (2007)
[296] B. Schnyder, D. Alliata, R. Kötz, and H. Siegenthaler, *Appl. Surf. Sci.*, **173** (2001) 221.
[297] J. M. Soler, A. M. Baro, N. Garcia, and H. Rohrer, *Phys. Rev. Lett.*, **57** (1986) 444.
[298] K. K. Cline, M. T. McDermott, and R. L. McCreery, *J. Phys. Chem.*, **98** (1994) 5314.
[299] J.-P. Randin and E. Yeager, *J. Electroanal. Chem.*, **36** (1972) 257.
[300] R. J. Rice and R. L. McCreery, *Anal. Chem.*, **61** (1989) 1637.
[301] K. Ray and R. L. McCreery, *Anal. Chem.*, **69** (1997) 4680.
[302] M. J. Esplandiu, H. Hagenstrom, and D. M. Kolb, *Langmuir*, **17** (2001) 828.
[303] K. Kopitzki, Einführung in die Festkörperphysik, Teubner Studienbücher: Physik, 1993.
[304] J. T. Grant and T. W. Haas, *Surf. Sci.*, **21** (1970) 76.
[305] C. Stampfl and M. Scheffler, *Phys. Rev. B*, **54** (1996) 2868.
[306] C. Stampfl, S. Schwegmann, H. Over, M. Scheffler, and G. Ertl, *Phys. Rev. Lett.*, **77** (1996) 3371.
[307] W. F. Lin, M. S. Zei, Y. D. Kim, H. Over, and G. Ertl, *J. Phys. Chem. B*, **104** (2000) 6040.
[308] M. Mavrikakis, J. Rempel, J. Greeley, L. B. Hansen, and J. K. Nørskov, *J. Chem. Phys.*, **117** (2002) 6737.
[309] CRC Handbook of Chemistry and Physics, 76th ed., CRC Press, New York, 1996.
[310] K. Tanaka and A. Sasahara, in Interfacial Electrochmistry (A. Wieckowski, ed.), Dekker, M., New York, 1999, p. 493.
[311] Q. Xu, U. Linke, R. Bujak, and T. Wandlowski, *Electrochim. Acta*, **54** (2009) 5509.
[312] A. J. Bard, R. Parsons, and J. Jordan, Standard Potentials in Aqueous Solutions, Marcel Dekker, New York, 1985.
[313] O. M. Magnussen, J. Hotlos, R. J. Nichols, D. M. Kolb, and R. J. Behm, *Phys. Rev. Lett.*, **64** (1990) 2929.
[314] N. Batina, T. Will, and D. M. Kolb, *Faraday Discuss.*, **94** (1992) 93.
[315] M. Cappadonia, U. Linke, K. M. Robinson, J. Schmidberger, and U. Stimming, *J. Electroanal. Chem.*, **405** (1996) 227.
[316] M. Wasberg, M. Hourani, and A. Wieckowski, *J. Electroanal. Chem.*, **278** (1990) 425.
[317] D. M. Anjos, M. A. Rigsby, and A. Wieckowski, *J. Electroanal. Chem.*, **639** (2010) 8.
[318] D. M. Kolb, Przasnys.M, and Gerische.H, *J. Electroanal. Chem.*, **54** (1974) 25.
[319] E. Bertel, G. Rosina, and F. P. Netzer, *Surf. Sci.*, **172** (1986) L515.
[320] H. B. Michaelson, *J. Appl. Phys.*, **48** (1977) 4729.
[321] D. Weingarth, Diploma Thesis, Technische Universität München, (München), 2009.
[322] J. Meier, J. Schiotz, P. Liu, J. K. Norskov, and U. Stimming, *Chem. Phys. Lett.*, **390** (2004) 440.
[323] T. Imokawa, K.-J. Williams, and G. Denuault, *Anal. Chem.*, **78** (2006) 265.

[324] T. B. Flanagan and F. A. Lewis, *Trans. Faraday Soc.*, **55** (1959) 1409.
[325] S. Gsell, M. Fischer, M. Schreck, and B. Stritzker, *J. Cryst. Growth* **311** (2009) 3731.
[326] A. Racz, P. Bele, C. Cremers, and U. Stimming, *J. Appl. Electrochem.*, **37** (2007) 1455.
[327] C. L. Green and A. Kucernak, *J. Phys. Chem. B*, **106** (2002) 1036.
[328] A. R. Kucernak, P. B. Chowdhury, C. P. Wilde, G. H. Kelsall, Y. Y. Zhu, and D. E. Williams, *Electrochim. Acta,* **45** (2000) 4483.
[329] R. G. Bates, D. B. Cater, G. J. Hills, D. J. G. Ives, G. J. Janz, R. W. Laity, I. A. Silver, and F. R. Smith, Reference Electrodes - Theory and Practice, Academic Press Inc., New York and London, 1961.
[330] J. Greeley and J. K. Nørskov, *Surf. Sci.*, **601** (2007) 1590.
[331] E. Santos and W. Schmickler, *Angew. Chem.,Int. Ed.*, **46** (2007) 8262.
[332] E. Santos and W. Schmickler, *ChemPhysChem*, **7** (2006) 2282.
[333] C. Camus, Diploma Thesis, Technische Universität München, (München), 2004.
[334] O. Paschos and U. Stimming, private communication.
[335] G. Karlberg and J. K. Norskov, private communications.
[336] T. Löffler, R. Bussar, E. Drbalkova, P. Janderka, and H. Baltruschat, *Electrochim. Acta,* **48** (2003) 3829.
[337] J. Greeley, T. F. Jaramillo, J. Bonde, I. B. Chorkendorff, and J. K. Norskov, *Nat. Mater.*, **5** (2006) 909.
[338] J. L. Zhang, M. B. Vukmirovic, Y. Xu, M. Mavrikakis, and R. R. Adzic, *Angew. Chem., Int. Ed.,* **44** (2005) 2132.
[339] J. Greeley and J. K. Nørskov, *J. Phys. Chem. C*, **113** (2009) 4932.
[340] M. Watanabe, H. Sei, and P. Stonehart, *J. Electroanal. Chem.*, **261** (1989) 375.
[341] H. Yano, J. Inukai, H. Uchida, M. Watanabe, P. K. Babu, T. Kobayashi, J. H. Chung, E. Oldfield, and A. Wieckowski, *Phys. Chem. Chem. Phys.*, **8** (2006) 4932.
[342] J. Kim, C. Jung, C. K. Rhee and T. H. Lim, *Langmuir*, **23** (2007) 10831.
[343] W. Tang, S. Jayaraman, T. F. Jaramillo, G. D. Stucky, and E. W. McFarland, *J. Phys. Chem. C,* **113** (2009) 5014.

Appendix

A1 Abbreviations and Symbols

a		Lattice Constant [m]
$a_i(I)$		Activity of ion species i in liquid phase I
A		Electrochemically active surface area [cm^2]
AFM		Atomic Force Microscope
Au		Gold
c		Concentration [Ml^{-1}]
C_d		Differential Capacitance [Fcm^{-2}]
C_D		Capacitance of Diffuse Layer [Fcm^{-2}]
C_H		Capacitance of Helmholtz Layer [Fcm^{-2}]
CE		Counter Electrode
CV		Cyclic Voltammetry
Cu		Copper
D		Diffusion Coefficient [m^{2}s^{-1}]
DOS		Density of States [(eV cm^3)$^{-1}$]
e		Electron Charge (e = 1.602*10^{-19}C)
E_F		Fermi Level [eV]
EC		Electrochemical / Electrochemistry
EC-SPM		Electrochemical Scanning Probe Microscope
EC-STM		Electrochemical Scanning Tunnelling Microscope
EDL		Electrochemical Double Layer
F		Faraday Constant (F = 96485 C mol^{-1})
f_{H2}		Fugacity of Hydrogen
ΔG^0		Standard Gibbs Free Energy Change [kJM^{-1}]
h		Planck Constant (h = 4.136*10^{-15} eV s)
$\hbar$		Reduced Planck Constant ($\hbar$ = 6.582*10^{-16}eV s)
h		Step Height [m]
H		Hydrogen
HOPG		Highly Oriented Pyrolytic Graphite
I		Current [A]
I_{Tun}		Tunnelling Current [A]
I_{Tip}		Tip Current [A]
IHP		Inner Helmholtz Plane
j		Current Densitry [Acm^{-2}]
k		Rate Constant [s^{-1}]
k_B		Boltzmann Constant (k_B = 8.617*10^{-5} eV K^{-1})
LDOS		Local Density of States [(eVcm^3)$^{-1}$]
m		Mole Number
M		Mole
ML		Monolayer
n		Number of transferred electrons

NHE	Normal Hydrogen Electrode
OCP	Open Circuit Potential
OHP	Outer Helmholtz Plane
Pd	Palladium
Pt	Platinum
PZC	Point of Zero Charge [V]
Q	Charge [C]
R	Gas Constant (R = 8.314 J(mol K)$^{-1}$
RE	Reference Electrode
SECPM	Scanning Electrochemical Potential Microscope
SEM	Scanning Electron Microscope
SPM	Scanning Probe Microscope
STM	Scanning Tunnelling Microscope
t	Time [s]
T	Temperature [K]
U	Potential [V]
U_{Bias}	Bias Voltage [V]
U_S	Potential of the sample in EC-SPM
U_{tip}	Potential of the tip in EC-STM
U_0	Redox Potential
U_{00}	Standard Potential
WE	Working Electrode
$z_i(I)$	Charge of ion species i in liquid phase I
α	Transfer coefficient
α_a	Anodic transfer coefficient
α_c	Cathodic transfer coefficient
ε	Dielectric constant [As(Vm)$^{-1}$]
ε_0	Permittivity of free space (ε_0 = 8.854*10^{-12} As(Vm)$^{-1}$)
η	Overpotential [V]
κ^{-1}	Debye Hückel Length [m]
ϕ	Electrostatic potential of electrode [V]
φ_0	Potential of the electrode [V]
ϕ_S	Work Function of sample [eV]
Θ	coverage [ML]
ν	Sweep Rate [Vs^{-1}]

A2 Publications

1. H.Wolfschmidt, R.Bußar, U.Stimming; "Charge transfer reactions at nanostructured Au(111) surfaces: influence of the substrate material on electrocatalytic activity"; *Journal of Physics: Condensed Matters* **20** (2008) 374127.

2. H.Wolfschmidt, D.Weingarth, U.Stimming; "Enhanced reactivity for hydrogen reactions at Pt nanoislands on Au(111)"; *ChemPhysChem* **11** (2010), 1533.

3. H.Wolfschmidt, C.Baier, S.Gsell, M.Fischer, M.Schreck, U.Stimming; "STM, SECPM, AFM and Electrochemistry on Single Crystalline Surfaces"; *Materials* **3** (2010), 4196

4. H.Wolfschmidt, O.Paschos, U.Stimming; "Hydrogen Reactions on Nanostructured Surfaces"; in: A.Wieckowski, J.K.Nørskov (Eds.), *Fuel Cell Science: Theory, Fundamentals, and Biocatalysis*, New York, 2010, 1

5. H. Wolfschmidt, M. Rzepka, U. Stimming; "Wasserstoff – Energieträger der Zukunft" / "Hydrogen as Future Energy Carrier"; *Konferenzbeitrag zum 32. Internationalen Wiener Motorensymposium, VDI Fortschrittsberichte* (2011)

6. P. Quiano, E. Santos, H. Wolfschmidt, U. Stimming, W. Schmickler; "Combined Experimental and Theoretical Studies towards Hydrogen Reactions on Pt and Pd Nanoparticles on Au(111)" *Catalysis Today* **177** (2011) 55

7. M.E. Björketun, G.S. Karlberg, H. Wolfschmidt, U. Stimming, J. Rossmeisl, I. Chorkendorff, and J.K. Nørskov; "Hydrogen evolution on Au(111) covered with submonolayers of Pd" *Physical Review B* **84** (2011) 045407

8. H. Wolfschmidt, U. Stimming; "Hydrogen, Oxygen and Methanol Reactions on Nanostructured Pt/Au(111) Surfaces", in preparation

9. G. Wang, Y, Wang, P. Zhang, X. Yang, H. Wolfschmidt, U. Stimming, J. Li; "Combined Experimental and Theoretical Investigations on the Hydrogen Adsoprtion and Spill-Over on Pd_6/Au(111) Cluster", *Journal of Physical Chemistry C*, in preparation

i want morebooks!

Buy your books fast and straightforward online - at one of world's fastest growing online book stores! Environmentally sound due to Print-on-Demand technologies.

Buy your books online at
www.get-morebooks.com

Kaufen Sie Ihre Bücher schnell und unkompliziert online – auf einer der am schnellsten wachsenden Buchhandelsplattformen weltweit! Dank Print-On-Demand umwelt- und ressourcenschonend produziert.

Bücher schneller online kaufen
www.morebooks.de

 VDM Verlagsservicegesellschaft mbH
Heinrich-Böcking-Str. 6-8   Telefon: +49 681 3720 174   info@vdm-vsg.de
D - 66121 Saarbrücken   Telefax: +49 681 3720 1749   www.vdm-vsg.de

Printed by Books on Demand GmbH, Norderstedt / Germany